Brave N

'I am now a convert. Aaron sums up all the ~~~~~~~~~~
to create teams and companies that maximize their potential by
decentralizing their power – a once-idealist notion that is now
possible and essential. For a book that might start a revolution, it's
surprisingly practical and undogmatic. There's no fluff – it's all meat,
and real news. I could think of dozens of people I know who I now
want to read and study it'
**Kevin Kelly, author of *The Inevitable*, and co-founder, *Wired*
magazine**

'This book is a breath of fresh air'
**Adam Grant, *New York Times* bestselling author of *Give and Take*,
Originals, and *Option B* with Sheryl Sandberg**

'The one-size-fits-all monoculture is a thing of the past'
**Susan Cain, author of *Quiet* and *Quiet Power*, curator of Quiet
Revolution**

'Complexity conscious. People positive. These words from Aaron's
book keep ringing in my ears as I work. He reframes the way we see
companies and the way we see each other inside them, and gives us
practical tools to transform ourselves and our environments. Whether
you lead a company, work at one or imagine yourself founding one
someday, read this book immediately and start an evolution'
Miguel McKelvey, co-founder, WeWork

'If you're trying to create a world-changing culture, reading *Brave
New Work* should be your next move. Aaron's simple, counter-
intuitive approach will help you get out of your own way, eliminate
bureaucracy and awaken the humanity within'
Scott Harrison, founder, charity: water, and author of *Thirst*

'Human beings can't thrive in a work culture that uses burnout and
"being always on" as proxies for dedication and success. In *Brave
New Work*, Aaron Dignan shows us that, in fact, workplaces that
empower people to take care of themselves are far more likely to
deliver sustainable performance and happiness'
Arianna Huffington, founder and CEO, Thrive Global

'We tend to look for answers by looking reflectively backwards – it's what we've all been taught in school. But Dignan insists that the "best practices" of the past no longer work because the bureaucracies of existing organizations have been defeated by new technologies. Instead we can only find those answers by "living in the now" the way a new breed of organization is already beginning to master'
John Maeda, Head of Computational Design and Inclusion, Automattic

'I really never believed in any of this organizational stuff until I saw Aaron Dignan at work. He can help almost any dysfunctional group find common purpose, discern the simple patterns underlying the most complex situations, and guide wayward organizations back to their core values. Most impressively, he can translate all that into language even a businessperson can understand and enjoy'
Douglas Rushkoff, author of *Team Human* and *Present Shock*

'This book will teach you to wrestle and win against workplace bureaucracy. Aaron cuts to the core of what makes teams successful by realigning hearts, minds, and egos. He always sparks better outcomes, and his book will be just the spark you need to get started'
Beth Comstock, author of *Imagine It Forward*, and former Vice Chair, GE

ABOUT THE AUTHOR

Aaron Dignan is the founder of The Ready, an organization design and transformation firm that helps institutions like Johnson & Johnson, Charles Schwab, Kaplan, Microsoft, Lloyds Bank, Citibank, Edelman, Airbnb, Cooper Hewitt, Smithsonian Design Museum and charity: water change the way they work. He is a co-founder of responsive. org, an angel investor in purpose-driven startups and a friend to misfit toys. He lives in Denver with his wife and son.

BRAVE NEW WORK

Are You Ready to Reinvent
Your Organization?

Aaron Dignan

PENGUIN BOOKS

PENGUIN BOOKS

UK | USA | Canada | Ireland | Australia
India | New Zealand | South Africa

Penguin Books is part of the Penguin Random House group of companies
whose addresses can be found at global.penguinrandomhouse.com.

First published in the United States of America by Portfolio Penguin 2019
First published in Great Britain by Penguin Business 2019
This edition published with a new foreword in Penguin Books 2023
001

Printed and bound in Great Britain by Clays Ltd, Elcograf S.p.A.

The authorized representative in the EEA is Penguin Random House Ireland,
Morrison Chambers, 32 Nassau Street, Dublin D02 YH68

A CIP catalogue record for this book is available from the British Library

ISBN: 978-0-241-99873-1

www.greenpenguin.co.uk

Penguin Random House is committed to a
sustainable future for our business, our readers
and our planet. This book is made from Forest
Stewardship Council® certified paper.

For Huxley

May you inherit a world in which everyone
finds fulfillment and prosperity at work.
And if you don't, may you help build one.

Contents

People who like this sort of thing will find this the sort of thing they like.

—**Artemus Ward**

Everything that needs to be said has already been said. But since no one was listening, everything must be said again.

—**André Gide**

Foreword

In 1941, Walt Disney Productions released *Dumbo*, an animated feature about an elephant who is criticized for his big ears, until he learns that he can use them to fly. But first, he needs to believe in himself. To that end, he is given a "magic" feather, which gives him the confidence to take the leap. Eventually, he learns he doesn't need the feather at all.

For most of my career, the hardest part about my job has been finding that magic feather for my clients. What will give them the confidence to change the way they work? What will help them believe that change is possible? This is harder than it looks. It's far easier to teach new ways of working to someone who has already found their confidence—someone already committed to the principles this book espouses. Since *Brave New Work* was first published, I've met thousands of leaders and teams who are ready to fly or are already flying—yet at the same time, I've met many more thousands of people who still don't have the confidence to jump.

In the end, I found our magic feather where I least expected it: on the heels of a global health crisis. In early 2020, something happened that would change work forever. The coronavirus pandemic swept the globe, forcing billions of people to quarantine and, wherever possible, work from home. Nearly every corporation on Earth made a sudden

transition to remote or hybrid work. Digital transformations that were meant to take *years* were rolled out in a matter of *weeks*. Office buildings, once great gathering places, stood empty—tributes to a bureaucratic world gone by.

Of course, the remote and hybrid work that many of us experienced didn't amount to a true revolution. We could have used this opportunity to really reinvent the way we work. Instead, we traded a day of meetings at the office for back-to-back Zoom meetings. Nevertheless, this was a *huge* change, and like anything new and uncertain, it was scary. Then, to our great surprise, it worked. It wasn't business as usual, but it was still business. And after a while, it became clear that for most organizations, things were going to be okay.

The pandemic was a tragedy, full stop, with few silver linings. But the transition to remote and hybrid work contained a secret gift that almost no one noticed. It was a *shared* experience of rapid change on a global scale. It was something almost every working adult could point to and say, "We tried that, and we did okay." And today, I can ask almost any team in the world a simple question: "Do you remember when you uprooted your way of working in a matter of days . . . and you survived?" They do remember. And together, we may well wonder, "If we pulled that off, what *else* is possible?"

Since the pandemic, we've heard from so many former clients, colleagues, and friends about how working with the principles outlined in this book—autonomy, transparency, and consent—set them up to succeed in incredibly trying times. They didn't need to mandate change from the top—because their systems were already changing continuously. They weren't perfect. But they were able to navigate challenges that truly stifled other teams. I wish that everyone could experience what it feels like to watch an organization come together around a problem and dance with it like a flock of birds.

I don't know what's next for the world of work. No one does. But I'm sure that we'll face another wave of change. And then another. We can keep getting knocked down. Or we can grab our magic feather, take the leap, and fly. This book isn't the answer, it's the question. The only one that matters: what's stopping you?

Part One

THE FUTURE OF WORK

The beginning is the most important part of the work.

—Plato

WE WERE PACKED INTO THE BACK OF A BLACK CAR, ON OUR
way to a celebratory dinner. The energy was electric. My clients and I
had just spent the last eight hours talking about the things teams never
get a chance to talk about, starting with the ultimate question, "What's
stopping you from doing the best work of your life?" Pent up for so long,
the possibility of actually addressing what held them back had awakened
something deep within this team. The banter from the room had contin-
ued into the hall, the elevator, and now the car. "What about our monthly
strategy review?" one of them said. "Does anybody actually get value out
of that thing?" Those words hung in the air for a few seconds before
everyone snapped back, "No!" The team leader turned around from the
front seat and looked everyone in the eye. "If you don't need it . . . I guess
I don't need it either."

Well, that did it. From what I could gather in the commotion, this
leadership team and the people that reported to them spent an inordi-
nate amount of time each month preparing for and attending what

essentially amounted to a glorified update meeting. I took the bait. "How much time, ballpark, would you say you spend on that meeting?" They started doing the math. First they had the cost of the meeting itself, which was more than three hours long and included almost forty people. Then there was preparation time for those senior leaders—they had to be sure they knew what they were talking about when questions came up. Add to that the time their teams spent preparing the material—hundreds of pages of PowerPoint, many of which routinely didn't get used or seen. And so on. They rattled it off and I tallied it up. By the time we got to the restaurant we had a good back-of-the-napkin estimate. The annual cost of the meeting? Close to $3 million. Their minds were blown. "We are spending $3 million to have this shitty meeting?! What should we do?" "Well," I said, "why not just cancel it and see what happens?" You could almost see the gears turning in their heads. If meetings could change, *what else* was up for grabs—budgets, approvals, structure? This was a team about to reinvent their way of working.

WORK ISN'T WORKING

No matter where I go in the world—and my work has taken me to fifteen countries and five continents—I meet leaders and teams who are frustrated. We are all confronted with the fact that the scale and bureaucracy that once made our organizations strong are liabilities in this era of constant change. We are beset on all sides by pressure—to grow, to deliver, to execute at all costs, and to do so with our arms tied behind our backs. We are being asked to invent the future, but to do so inside a culture of work that is deeply broken.

We don't have enough time to do our work, but we pack our days with endless meetings. We don't have the information we need, but we are buried in emails, documents, and data. We want speed and innovation, but we run from risk and inhibit our best people. We claim to work

in teams, but we don't really trust one another. We know the way we're working isn't working, but we can't imagine an alternative. We long for change, but we don't know how to get it.

Today we face an array of systemic challenges—in our economy, our government, our environment—all stemming from our inability to change. We are addicted, in spite of ourselves, to the siren song of bureaucracy. Gratuitous hierarchies, plans, budgets, and controls abound. But they aren't working like they used to. In the face of complexity, our *Legacy Organizations*—the traditional institutions that make up much of the modern world—are failing us, and we know it. Yet we do nothing, paralyzed by the fear of losing whatever control we have left.

SIMPLE SABOTAGE

Think back on your career, whether that's a few years or a few decades. Think about the things that have frustrated you. What has held you back. What do you wish you could change. Now read the following instructions and see if you recognize them. Have you ever seen a colleague behave this way?

1. Insist on doing everything through "channels." Never permit shortcuts to be taken in order to expedite decisions.
2. Make "speeches." Talk as frequently as possible and at great length. Illustrate your "points" by long anecdotes and accounts of personal experiences.
3. When possible, refer all matters to committees for "further study and consideration." Attempt to make the committees as large as possible—never less than five.
4. Bring up irrelevant issues as frequently as possible.
5. Haggle over the precise wordings of communications, minutes, resolutions.

6. Refer back to matters decided upon at the last meeting and attempt to reopen the question of the advisability of that decision.

7. Advocate "caution." Be "reasonable" and avoid haste which might result in embarrassments or difficulties later on.

8. Be worried about the propriety of any decision—raise the question of whether such action as is contemplated lies within the jurisdiction of the group or whether it might conflict with the policy of some higher echelon.

9. When training new workers, give incomplete or misleading instructions.

10. Hold conferences when there is more critical work to be done.

11. Multiply the procedures and clearances involved in issuing instructions, paychecks, and so on. See that three people have to approve everything where one would do.

12. Apply all regulations to the last letter.

Are you laughing? Most of the leaders I know can't make it through that list without howling. They've seen it all before. Hell, they've seen it *this week*. Now, perhaps you've spent your career in startups and none of that struck a chord for you. That's good. But read on, because this is a cautionary tale—a glimpse of what's to come if you're lucky enough to scale.

The obvious question: who wrote these? One plausible answer is that these are behaviors I've observed over many years of working with large clients. It's a field report. A sad bureaucratic ethnography. And while that is certainly possible—I have seen all this and then some—the actual source is far more unexpected and profound. In fact, it's almost unbelievable.

In 1944, at the height of World War II, William J. Donovan was the director of the United States Office of Strategic Services (the precursor to the CIA). He was looking for ways to undermine and destabilize enemy states, particularly those that figured prominently in the war. To

that end, he commissioned a new field manual to be developed by the agency. The manual would be provided to ordinary citizens inside enemy territory who were sympathetic to the Allies. It would have a singular focus: helping these citizens carry out acts of "simple sabotage" that would destabilize their own communities and businesses. When completed, the so-called *Simple Sabotage Field Manual* was wide-ranging. It covered everything from damaging buildings and infrastructure to interrupting supply lines. But there at the end was a little section focused on disrupting *day-to-day business operations*.

As you may have deduced by now, the instructions in the list on the previous page—the ones that sound so eerily familiar—were copied verbatim from that manual, written in 1944. And yet here we are. We see our organizations in those instructions. We see our colleagues. We see ourselves. Somehow, in less than a lifetime, modern work has become indistinguishable from sabotage.

Early on in almost every client relationship, a leader will pull me aside and ask, "Honestly, are your other clients as f@#%ed up as we are?" Whereas the saltiness of the language may vary, the theme remains constant. Their question betrays a paradox at the heart of our struggle with work. Either everyone else has it mostly figured out, and we're alone in our bureaucratic melancholy, or we're surrounded by firms facing the same challenges, and the world is deeply dysfunctional. Both would be bittersweet. Of course, I simply tell the truth. "We see this all the time. Everyone is in the same boat." They visibly relax. Misery loves company.

EVERYTHING HAS CHANGED, EXCEPT MANAGEMENT

For five years, I've opened nearly every speech I've given with a picture of an org chart. I ask the audience what year it's from. Here's the thing: nobody has any idea. In every country, in every setting, I hear dates

ranging from 1800 to yesterday. People call out haphazard guesses, laughing under their breath, because they know the truth: it could have been made at *any* time. It looks exactly like the org chart in their organization. And I'll bet it looks exactly like your org chart too.

The one featured here happens to be more than a hundred years old. It's a railroad org chart, published in 1910 as the Operating Organization of the Union Pacific and Southern Pacific Systems. What's really shocking about our inability to "carbon-date" this document is how easy it would be if it were a picture of almost anything else. If I showed you a house, a car, a dress, or a phone from 1910 and asked you whether it was modern or antique, you'd have a pretty good idea. Because almost everything has changed. But not management. Not really. Information flows up. Decisions flow down. A place for everyone, and everyone in their place.

Somehow, amid a period of relentless innovation, including the internet, mobile computing, autonomous vehicles, artificial intelligence, and rockets to space that can land themselves, the way we come together as human beings to solve problems and invent our future has stayed remarkably constant. Which means one of two things is true: either

we've perfected the way we organize, and we should all submit to the power of the pyramid, or we're stuck in a Gordian knot of our own design, unable to break free and realize a better way.

HOW TO CROSS THE STREET

What's hiding inside each of our organizations is a set of assumptions that we rarely notice or reconsider. We've inherited them from those who came before us. These assumptions, and the practices they inspire, are kind of like an *operating system* (OS), running silently in the background, the foundation upon which everything else is built.

Operating systems are all around us. Take intersections. Two roads crossing present a deceptively simple challenge: how do we prevent cars from hitting one another, while maintaining the maximum flow of traffic?

One of the most popular solutions to this problem is the signal-controlled intersection. With an estimated 311,000 signals in the United States

alone, nearly everyone on Earth is familiar with this approach. Now, if we consider this as an operating system, we can ask, What are the assumptions behind it?

> People cannot be trusted to manage the intersection on their own. They need to be told what to do.
> Complex problems must be managed with elaborate rules and technology, in the form of cables, lights, switches, and control centers, programmed to optimize the flow of traffic.
> We need a plan for every possible scenario, with multicolored signals, arrows, the ability to switch from solid to flashing lights, and so on.

Less prevalent, but still well known, is the roundabout. Here cars enter and exit a shared circle that connects all four directions of travel. This too represents an operating system—one that holds different assumptions about people and the problem.

> People can be trusted, and will trust one another, to use judgment and do the right thing.

> Complex problems can be managed with simple rules and agreements that leave room for judgment: give the right-of-way to vehicles already in the circle and go with the flow of traffic.
> Many scenarios will unfold in the roundabout, but social coordination will be sufficient to handle them.

What stands out for you about these two approaches? Did you notice that the signal-controlled intersection doesn't require much thinking? Only compliance. The roundabout, on the other hand, requires drivers to be present and responsible for their own safety and the safety of others. One allows for a surreptitious text message or two while we wait. The other one keeps things moving. One has a huge apparatus behind it, with control stations and staff monitoring the situation constantly. The other is left to its own devices.

We have precious few roundabouts in the United States, about one for every 1,118 intersections. So you would assume that the signal-controlled intersection is superior. But which operating system actually produces the best results? Let's test our intuition:

> Which one is safer?
> Which one has higher throughput?
> Which one is cheaper to build and maintain?
> Which one works better when the power goes out?

The answer to all of the above is the roundabout. They reduce injury collisions by 75 percent and fatal collisions by 90 percent. They result in an 89 percent reduction in delays. They are $5,000 to $10,000 cheaper to maintain each year. And of course, they work normally during power outages. And yet which one are we more comfortable with? The signal-controlled intersection. Surprising, right? We confuse popularity with quality.

These two approaches are a surprisingly good metaphor for what's happening in the world of work. The way of working that we know best isn't serving us, but it's hard to trust the alternative. Our organizational operating system—the practices, policies, processes, procedures, rituals, and norms that shape our day-to-day reality—is so prevalent we almost take it for granted. But if you've ever stopped to wonder why we need managers, or budgets, or performance reviews, you were questioning the operating system—the Legacy OS—*without even knowing it.*

> There's no word for accountability in Finnish. Accountability is something that is left when responsibility has been subtracted.
>
> —Pasi Sahlberg

If the assumptions baked into our OS are invalid, then no amount of commitment or resolve is going to make things better. We are addicted to the idea that the world can be predicted and controlled—that our stoplights are the only way to keep things in check. But when you view the world that way, today's uncertainty and volatility become triggers for retreating to what has worked in the past. We just need to *hire more capable leaders*. We just need to *squeeze out a little more efficiency and growth*. We just need to *reorganize*. . . . But we know better. The real barrier to progress in the twenty-first century is us.

THE FUTURE OF WORK

What if your organization could run itself? What if your corporation, your startup, your restaurant, your school, or your church were able to get better every day, without you having to move mountains to make it so? What if you could stop giving orders? Stop checking in to see how things are going? Stop obsessing over your budget, your plan, your next quarter? This is not only possible; it is *already happening* inside organizations around the world.

When it comes to how they build teams, manage projects, make decisions, share information, set targets, review performance, and set compensation, their approach is not a signal-controlled intersection but a roundabout. They use purpose, transparency, and reputation to create cultures of freedom and responsibility. They are intentional but full of serendipity. They are decentralized but coherent. Above all, these firms are *People Positive* and *Complexity Conscious*—two foundational mindsets that we'll explore in detail.

I refer to these iconoclasts as *Evolutionary Organizations* because they use these mindsets to collectively and continuously improve their shared OS. This isn't just a way of being; it's a way of *doing*. It is a *practice*. And whether you're leading a team of ten or ten thousand, improving *your* OS is the single most powerful thing you can do. That is the promise of this book.

It was written for leaders, executives, entrepreneurs, managers, philanthropists, elected officials, community organizers, superintendents— anyone who needs a team of people to achieve their collective potential. And that potential is vast: Such as the Swedish bank Handelsbanken outperforming its competition for decades without a traditional budget. Or nurses working in teams with no managers at Buurtzorg. Or employees writing their own job descriptions and setting their own salaries at Morning Star. As we go, we'll meet organizations that have done each of these things and many more. And you'll see that it's not chaos that produces these unlikely achievements but consciousness. Before we're done you'll wonder why anyone would want to keep working in captivity.

But first, a preview of things to come. In the remainder of Part One, we'll find out how our way of working came to be—originating on the factory floors of the Industrial Revolution. We'll see how and why it's failing us as we approach unprecedented scale and complexity. And we'll lay the foundation for a new way of thinking about people and organizations that trades the illusion of control for something far better.

In Part Two we'll learn about the principles and practices of Evolutionary Organizations through a tour of the *OS Canvas*, a tool my colleagues and I invented to help teams see just how embedded and interconnected our way of working really is. The canvas highlights twelve domains where we most urgently need to question and reinvent our approach: purpose, authority, structure, strategy, resources, innovation, workflow, meetings, information, membership, mastery, and compensation. I won't pretend there's a one-size-fits-all playbook for Evolutionary Organizations here. But that's good news. Because while we may share a set of principles, our practices will be shaped by the culture, context, and characters that surround us. And if you're worried about the fact that you don't call all the shots at work, don't be. Regardless of rank, here you'll find bite-sized ideas to change your team for the better.

In Part Three we'll tackle the most difficult topic of all: how to change. If you've been burned by change efforts that have overpromised and underdelivered, you're not alone. I'll present a counterintuitive yet straightforward approach to change that honors the inherent complexity in your organization. We'll learn how to turn tension and frustration into safe-to-try experiments that move teams forward. I'll suggest guidelines and activities for early meetings and moments in your team's change process. And I'll share stories and lessons from transformation efforts that struggled and those that went well beyond our expectations.

Finally, we'll take a moment to imagine a world in which we get this right, and organizations everywhere create fulfillment and abundance. The foundations of a new economy are slowly forming, and I'll show you where to look to see them taking shape. In the end, you'll have everything you need to step confidently into the future of work.

HOW I GOT HERE

In 2007 I was part of the founding team of a company that created digital strategies for some of the biggest brands in the world. A few years later I became the CEO. While I thought of myself as progressive, in truth I managed the company in a somewhat traditional way. I weighed in on all the hiring and firing decisions. I gave annual performance reviews and set compensation. I kept the firm's finances and salaries under lock and key. I told people what to work on, and how to do it well. I *designed* the organization as *I* thought it should be.

Now, to be fair, our culture was more permissive and flexible than most. We put new people in the deep end with important clients and let them figure out how to swim. We asked people for their opinions and answered (most of) their questions. And I truly believed my employees were operating at peak potential. Our firm had a patina of democracy about it. But when it came to the important decisions—strategy, design, brand, culture—I was often the one making the call.

And if I'm being honest, we did pretty well. We grew quickly and hired some amazing people—some of the best I've ever worked with. Our clients were happy. Our reputation was strong. From the outside it looked like success.

There was only one problem: it was exhausting. As it turns out, controlling everything is not really sustainable. And even when it works, it's unfulfilling. I would lie in bed night after night, sometimes at two or three in the morning after working a sixteen-hour day, with the same thought stuck in my head: this can't be the best way to run a business. Slowly the realization hit me. If I truly was the hero leader, the single point of failure, then what had I really built? A fragile system. In my quest for perfection I had limited my colleagues' ability to shape and grow the firm, and it was holding us back from our true potential.

At the same time, I was getting a front-row seat to how things were done in some of the largest companies in the world. And honestly, things didn't look good. As far as I could tell, big always translated to bad. Bigger companies were slower, less innovative, and less human places. People there were constantly wearing masks that hid their true selves, trading the long term for the short term, and actively disempowering their colleagues—it was everything I didn't want. But if we were to grow, wasn't that our future? That seemed to be the only option on the table: trade freedom, speed, and humanity for money and scale. I could feel my definition of success changing, but to what I didn't know.

I had questions and went looking for answers. Was there a way to run a company that didn't end in a bureaucratic mess? Was agility at scale even possible? Was growth a goal or a by-product? My walkabout led me to many unexpected sources of inspiration. I found complex adaptive systems—such as cities, ants, flocks of birds, schools of fish, immune systems, and brains—that solved problems without a leader, adapting to a wide variety of situations and challenges. I learned about radical concepts such as biomimicry and emergence, resilience and antifragility. And I began to discover a hidden-in-plain-sight collective of people and organizations that were doing things differently, some since before I was born. They represented a way of working known as *self-management*, characterized by a lack of traditional hierarchy and bureaucracy. These Evolutionary Organizations had been out there all along, but somehow popular business culture had managed to ignore their stories or rationalize how they didn't apply to the rest of us. How many people have ever heard about the inner workings of a pioneer such as W. L. Gore, the makers of Gore-Tex, or FAVI, a successful brass auto parts foundry that has been leveraging decentralization and distributed authority since 1983? And yet there were dozens and dozens of stories like these, both new and old. Each with its own contribution to a growing body of knowledge about how to bring humanity, vitality, and adaptivity back to work.

So I made a decision. We were going to change our OS. Self-management or bust. Easier said than done. I had to change my identity from a charismatic control freak who thought his job was to ensure perfect execution to someone who could let go and let come. But before I could do that I had to get everyone else on board. And that wasn't easy, because this change affected people in ways I couldn't even anticipate. Those who had previously wielded power had to find new ways to feel important and useful. Those who had lacked organizational power had to step into their new authority and take ownership of their working lives. This was hard work.

It didn't happen overnight. We tried a handful of new practices. Some worked, some didn't. We learned new principles and felt a little bit wiser. We argued. We debated. We became ethnographers of our own culture. One day we looked up and we were meeting differently. Deciding differently. Talking differently. Slowly but surely, we had become one of those companies that was not like the others. Were we done? Not at all. In fact, we'd realized what only those companies that really care about their way of working know—that you're never done. You're always learning. Always changing.

Twelve months later, when we looked back at our financial performance, we had delivered our best year ever. The funny thing was no one knew why. There wasn't any one thing we could point to and say, ah, *that* was the reason we grew, that was the reason we succeeded. It just . . . *happened*. And that's when I realized that we had gone from a place where all the best ideas were trapped in our heads to a place where they were free—where the genius of the firm was happening not inside people but *between* people. I knew I could never go back.

And I haven't. Years later when I started The Ready, a company that helps organizations change the way they work, I decided to put everything I had learned into practice from day one. And the results have been truly astounding. In just a few short years the firm has grown to serve

clients around the world and has helped tens of thousands change their working lives.

Meanwhile, I spent twelve months writing the book you're holding in your hands. Three to four days a week, every week, I stopped going into the office or out with clients. I wasn't interrupted. No emergency calls asking me to resolve a crisis. I just strolled into our office one summer day and told the team, "I'm writing a book to tell the story of our work. I'll be gone a lot while I work on this. Let me know if you need anything." They said, "Exciting! We'll miss you. Get lost." Consider for a moment how unusual that is. That the founder of a new business could spend 60–80 percent of their workweek *writing* without the whole thing imploding. At what normal company or startup can that happen?

Of course, I was not without my doubts. Could the culture sustain itself without me? Would performance go down? Would people stop showing up? Maybe. But I trusted the culture we'd built and jumped in with both feet. A year later, I'm happy to report that everything worked out. It wasn't effortless or without its challenges. After all, I held roles in the company that had to be filled by new people in new ways. But in the end, our organization rose to that challenge on its own. The company actually *grew*, if you can believe it, in more ways than one. All thanks to our amazing team.

Now, I'm sure some of you are wondering if I appointed an interim leader. Someone to "hold down the fort" or "make the key decisions"? Nope. Our way of working—our operating system—is built on distributed authority, consent, and agreements, so we don't have to rely on positional leaders. The average person at our company has more autonomy than a VP at a *Fortune* 500. Everyone has a credit card with no limit. Everyone has complete access to our finances. Even our *contractors*. When decisions arise, we have processes for making them quickly and

safely. When it comes to most things, I'm completely optional. And I've never been happier or prouder.

I don't say any of this to brag. There's nothing magical about what we've done here. Anyone can unlock this potential. But the secrets aren't out there in the halls of power. They're at the fringe. They're inside our organizations right now, in the minds of our people, trapped by decades of bureaucracy and habit. Your organization, and every other, has the innate ability to continually reinvent itself. You simply need to let it out.

A FINGER POINTING AT THE MOON

At this point you might be wondering how all this connects to the push for organizational agility that has been under way for the past few decades. It's an important question. The authors of the "Agile Manifesto" put forth a vision of a better way of developing software that prioritizes learning and agility over planning and control. The movement they spawned has inspired many people to think and work differently, well beyond the world of software. Now entire organizations are attempting to achieve "agility at scale," believing that it will be a remedy for all their problems.

Unfortunately, in their desire to *be* agile (with a lowercase *a*), many firms have attempted to *do* Agile (with an uppercase *A*). They have traded a goal for an orthodoxy—adopting the methods and certifications but not the theory behind them. And this is fatal. Because an agile workflow is not an operating system. The common practices of Agile development offer little advice to someone wondering about organizational structure, or human development, or compensation, or a thousand other questions that will emerge as you go. And even if it were comprehensive, you cannot cut and paste an OS any more than you can cut and paste a personality. *You* have to do the work.

The man who grasps
principles can successfully
select his own methods.
The man who tries
methods, ignoring
principles, is sure to
have trouble.

—Harrington Emerson

Japanese engineer Taiichi Ohno offered this advice to those seeking to learn his approach at Toyota: "Stop trying to borrow wisdom and think for yourself. Face your difficulties and think and think and think and solve your problems yourself. Suffering and difficulties provide opportunities to become better. Success is never giving up." Harsh but true. If you think certifying your project managers in Scrum is going to topple your bureaucracy, you're in for some major disappointment. Agility is a mindset, not a tool set. It's a piece of the puzzle, not the whole thing. It is necessary but not sufficient.

It turns out agility isn't an anomaly in this way. Many management innovations have emerged in the last half century, each promising to revolutionize work as we know it. Lean Manufacturing. Total Quality Management. ISO 9000. Six Sigma. Sociocracy. Holacracy. The Lean Startup. The list goes on and on. Each was, in its own way, a piece of an operating system. Some were misguided from the start. Others became perversions of themselves over time. And a few offered real wisdom that is yet to be fully realized.

Thich Nhat Hanh wrote in *Old Path White Clouds*, "A finger pointing at the moon is not the moon." When we focus too much on the method or the messenger, we lose sight of the deeper truth. The cases and stories I'll share in the pages ahead are fingers pointing at a better way of working. While we can draw inspiration from them, each of us must do the work of nurturing and evolving our own way.

But to do that, to truly take responsibility for our own future, we have to understand our past. Hierarchies, functions, plans, and budgets did not spring from the head of Zeus. Long before Agile development, before management consultants roamed the earth, a generation of orga-

nizational zealots saw a better way and willed it into existence. As a result of their efforts, we inherited a world where you can buy a McDonald's hamburger in 119 countries. But it's also a world where two out of three people are disengaged at work. And so before we dive headfirst into the principles and practices of Evolutionary Organizations, let's briefly take stock of how and where we went wrong.

THE ONE BEST WAY

The Legacy Organization was born more than a hundred years ago on a factory floor. Machinists then were an unruly bunch, employing wildly divergent customs and techniques even within the same facility. Novices learned on the job from the people around them, taking on their quirks, their tricks of the trade, and, of course, their pace. Given this artisanal flair, productivity was a moving target. No one—not machinists or managers—knew what was possible. Sure, they had a pretty good idea of how long it took to produce a specific part. But no one was asking how long it *should* take.

Enter Frederick Winslow Taylor. Promoted through the ranks at a young age, his innovation was to break the work down into its smallest possible components and find *the one best way* to complete each step. No task was too small or insignificant to provoke his scrutiny. This was a man who once made serious scientific study of *how to use a shovel*. His findings? Apparently every scoop should contain exactly 21.5 pounds of matter. If you think shoveling can be sufficiently done on instinct, you would not have found him sympathetic.

In Taylor's time, foot dragging and intentional inefficiency were so common there was a well-known term for it: soldiering. Taylor pulled no punches regarding this phenomenon. "Hardly a competent workman can be found in a large establishment . . . who does not devote a considerable portion of his time to studying just how slowly he can work and still

convince his employer that he is going at a good pace." Of course, this was the unintended consequence of a badly designed system. Employers were known to cut *piece rates*—the amount they paid workers for each completed part—when productivity and wages got too high, so workers limited their output to stave off any future rate cuts. Both parties found themselves in a standoff of predictable mediocrity.

To disrupt this dynamic, Taylor did what came naturally—an experiment. And this one flew in the face of decades of labor relations. He actually *raised* wages. But there were strings attached. He offered groups of workers as much as a 15–35 percent wage premium to do a particular job, but they had to do it *exactly* as he instructed. Later they could go back to doing it their way at the old pay or continue with the surplus pay and his exacting standards.

The question on his mind was simple: how much would it cost to get the average worker to give up their autonomy? The answer turned out to be: not that much. The implications of this mundane epiphany would change the world. In his detailed biography of Taylor, author Robert Kanigel spelled out this breakthrough. "And there it was—the Faustian bargain in embryonic form: You do it my way, by my standards, at the speed I mandate, and in so doing you will achieve a level of output I ordain, and I'll pay you handsomely for it, beyond anything you might have imagined. All you have to do is take orders, and give up your way of doing the job for mine."

> Technological progress has merely provided us with more efficient means for going backwards.
>
> —Aldous Huxley

Although he was well known in certain circles, Taylor didn't truly go mainstream until 1910. During a very public legal battle between the railroads and the Interstate Commerce Commission, future Supreme Court justice Louis D. Brandeis argued vigorously that the railroads could save more money by implementing the remarkable methods of "Scientific Management"

(a term he coined to describe Taylor's work) than by winning their case. Suddenly, efficiency was on everyone's mind.

On the heels of this newfound celebrity, Taylor published *Principles of Scientific Management*, which would become one of the bestselling business books of the decade. In it he offered four principles that would become the duties of a new class of managers:

First. They develop a science for each element of a man's work, which replaces the old rule-of-thumb method.

Second. They scientifically select and then train, teach, and develop the workman, whereas in the past he chose his own work and trained himself as best he could.

Third. They heartily cooperate with the men so as to insure all of the work being done in accordance with the principles of the science which has been developed.

Fourth. There is an almost equal division of the work and the responsibility between the management and the workmen. The management take over all work for which they are better fitted than the workmen, while in the past almost all of the work and the greater part of the responsibility were thrown upon the men.

Implicit in these principles was the idea that the *thinking* (the work for which management was "better fitted") and the *doing* (the hard labor) were to be separated once and for all. This is perhaps the most pronounced aspect of Taylor's legacy. Legions of MBAs followed in his footsteps, carrying with them the sacred duty of thinking for everyone else.

Following the publication of *Principles* in 1911, the world began to adopt Taylor's methods at an almost breakneck pace. And like any

guru, he was not without disciples and contemporaries. By this point in our story, a whole movement to tame the unruly world of work was under way.

Henri Fayol, a French mining executive, published *Administration Industrielle et Générale*, a theory of management that would come to be known as Fayolism. It contained principles such as *unity of direction*, suggesting that activities with the same objective should be directed by *one* manager using *one* plan for *one* common goal. And the *scalar chain*, which proposed that authority and communications should flow in a straight line from the *top to the bottom* of the organization. The boxes and lines of our modern org charts are the living embodiment of his principles.

Henry Gantt gave us the chart that bears his name—used to this day to illustrate dependencies within complicated projects and processes. But Gantt baked in a dangerous assumption: that the world can be predicted. In Gantt's repetitive world of manufacturing, this was tolerable, even helpful. But as his practices fanned out into knowledge work, they became a dangerous addiction. Instead of a means to an end, hitting the plan became an end unto itself.

And let's not forget James McKinsey, founder of consulting juggernaut McKinsey & Company, who completely upended the accounting paradigm while everyone else obsessed over productivity. In his view, budgeting was to be an expression of policy and strategy, flowing directly from the business plan. The budget should be used to evaluate performance, to see who beat their plan and who didn't. It would become another instrument of control. Like Taylor, McKinsey believed that great managers thought deeply about their processes and plans. He would often remark on this, in his trademark confrontational yet seductive tone. "Usually, I find that the executive who says he does not believe in an organization chart does not want to prepare one because he does not wish other people to know that he had not yet thought through his

organization properly. For the same reason, many men are opposed to budgets. They are unwilling for anyone to see how little they have thought about what they are going to do in future periods." See what he did there? If you disagree, it's because you haven't done the work.

These and so many more men and women made their mark. Frank and Lillian Gilbreth introduced time and motion studies, the precursor to ergonomics. Harrington Emerson fueled our fascination with efficiency. Hugo Münsterberg popularized the notion of vocational fit. Lyndall Urwick limited the span of control—how many people one leader can manage. Max Weber highlighted the value of rational-legal authority structures—the idea that positions and laws (not people) hold power. And of course, Henry Ford gave us his assembly line and the mass consumption that followed. Only Mary Parker Follett, the unsung mother of modern management, offered a more humanist opinion about how to realize our potential. Her ideas about reciprocal relationships (win-win) and noncoercive power (influence) were ahead of their time. Unfortunately, pieces of her work inadvertently led to the tangled matrix organizations we know and "love" today.

It would be unfair to paint these individuals as malicious in their intent. Each of them believed they were serving society by improving productivity and performance at a time when that was desperately needed. In so many ways they succeeded. It's quite safe to say our modern way of life wouldn't exist without them. And if we appreciate the reliability, accessibility, and convenience we now enjoy, we have them to thank, even as we struggle to transcend what they wrought.

The thinking separated from the doing. An addiction to prediction. A chain of command. And the convergence of it all in the form of an ironclad budget. Among the theories and contributions of these industrious few we can construct much of the tapestry of modern work. It's hard to argue their core tenets aren't still with us. We still tell people what to do (and how to think). We still demand detailed plans before

every initiative. We still focus on efficiency at the expense of effectiveness. We still use the budget as a weapon. You see, we can have all the foosball tables, fancy chairs, and free snack stations we want; we're still living in their world.

THE COST OF BUREAUCRACY

If Taylor worked at a multinational company today, he'd be flummoxed to find his principles metastasized into something truly *inefficient*. According to an employee at one federally owned corporation, toiletries have a *six-month lead time*. To get them, you must make a service request and get a Problem Evaluation Report. Then you need to get a work order, and a purchase order must be processed by the Procurement Engineering Group, who will then contact the vendor to ensure no changes have been made and the products meet regulatory requirements. And 180 days later, if you're lucky, you'll get your toilet paper.

We have a name for this phenomenon: bureaucracy. For most people, the term evokes a feeling of soul-crushing inefficiency and boredom. In contrast to Scientific Management's "one best way," it would seem that the aim of bureaucracy is to find the one *worst* way—that which wastes the most time and involves the most people and steps. The origin of the word itself is the unlikely pairing of *bureau* (French for desk) and *-kratia* (Greek for power or rule). So, in a very real sense, it means *management by desk*, which sounds about right.

London Business School professor Gary Hamel has taken up a crusade against bureaucracy by trying to put a value on our lost time and energy. After a broad workforce analysis, Hamel and his coauthor, Michele Zanini, claim that roughly half of the 23.8 million management roles in the United States are unnecessary. They found that a new wave of companies (including many featured in this book) have managed to cut their manager-to-employee ratio in half while keeping performance

up. It follows that if the market could leverage the methods of these pioneers, 12.5 million managers would be free to do something more productive.

Beyond the waste of management, U.S. workers spend a staggering 710 million hours per week on internal compliance activities such as budgeting and planning. That's 16 percent of our working lives. Yet, by some estimates, roughly half of all compliance activity isn't adding value. What equates to the work of 9 million people is being wasted each year on bureaucratic theater, and this show has no intermission.

Combined, we're talking about 21.5 million employees who aren't even scratching the surface of their potential. If they were doing value-added work at an adjusted average GDP per worker of $141,000, they'd conservatively add $3 trillion to the U.S. economy. Let's write that out and show it the respect it deserves: $3,000,000,000,000. And for those thinking globally, the research suggests there's another $5.4 trillion abroad. That's the cost of bureaucracy. It's a hidden debt—an organizational debt—that we pay interest on every day.

ORGANIZATIONAL DEBT

If your organization owes money, you know it. Financial debt is serious business. Fail to service it, and you'll find yourself bankrupt. If you develop software, you may also struggle with *technical debt*, the price you pay for taking shortcuts while coding. But there is another kind of debt that gets far less consideration, though it has the capacity to completely destroy your culture. It's called *organizational debt*.

Steve Blank, one of the pioneers of the lean startup movement, initially defined it as "all the people/culture compromises made to 'just get it done' in the early stages of a startup." Due to a lack of time, resources, or willingness to do the hard things, founders will put off developing fundamental programs such as employee onboarding or training. Or

they'll keep someone in a role long after they know it's not a fit. But org debt is so much bigger than that. It exists well beyond the domain of startups. In fact, I think it's highly correlated with organizational maturity and scale. I define organizational debt as *any structure or policy that no longer serves an organization*. Within that definition, we see it manifest in many different ways at many different times.

One of the most common sources of org debt is the knee-jerk reaction. Every time something goes wrong, we immediately jump to create a constraint that will prevent future mistakes. So we institute a new role, rule, or process. As a result, over the course of a decade or two, a one-step process becomes twenty steps. Or five different processes become entangled. Or ten different roles become approvers for a simple decision. And so it goes, on and on. This increase in complication creates more risk, so we create derivative roles and rules to manage it (they're called project management offices, or PMOs). Under the guise of creating order, we drift toward the *disorder* of a thousand stupid rules, leaving no ability to respond to the world as it unfolds.

Another source of org debt is change. In the world we live in, our roles, rules, and processes are at risk of becoming obsolete at any time. Some of them, while they may have been perfect for the context in which they were created, are now actively holding us back. And since we change jobs with increasing frequency, few among us were even here when they were cooked up in the first place. The ever-growing invite list for the monthly meeting that nobody likes? That's debt. Using last year's budget as a baseline for this year's budget? That's debt. A policy prohibiting employees from writing or speaking about their work on social media? That's debt. The question that must be asked is "What would we do if we were starting with a blank sheet of paper?" If the answer is anything other than what we've got, we have work to do.

The point is this: to avoid the pitfalls of organizational debt, we need constant and vigilant simplification. We need to create roles, rules, and

processes that are inherently agile—built to learn and change. Unfortunately, the very bureaucracy that created our org debt also stands in the way of addressing it. Org debt creates bureaucracy, and bureaucracy protects org debt. It's a tragic love affair.

GRACEFUL DEGRADATION

In spite of—and in part because of—bureaucracy, the world is doing better than ever. Since the publication of Taylor's *Principles* in 1911, the percentage of people living in extreme poverty has gone from 82.4 percent to just 9.6 percent. The number of people with a basic education has jumped to nearly nine in ten. Child mortality is one-eighth what it was. Democracy has become the dominant form of government on Earth. And the average global citizen has experienced a sixfold increase in prosperity. In so many ways we have been remarkably successful.

But under the surface something's not right. We can't quite put our finger on it, but work is harder than it used to be. Hours are long. Cycles are short. Wages are stagnant. Meaning is missing. The thrill is gone. Somehow, amid all our achievements, work has been failing us in a measurable way, but only a handful of people are seeing behind the curtain.

Among them are Richard Foster and Innosight, who have been studying *corporate longevity* within the S&P 500—a curated list of publicly traded companies that represents the U.S. stock market. Their research shows that in 1958, the average tenure of a company on that list was sixty-one years. But by 2016 that number had been reduced to twenty-four years. The data suggests this pattern will continue. At the current rate of churn, about half the list will be replaced in the next decade. And by 2027 the average tenure will shrink to just twelve years. This fits with broader research recently completed by the Santa Fe Institute in New Mexico. In its analysis of more than 25,000 companies, it found that the half-life of *all firms* was roughly 10.5 years. Big business,

Average Company Lifespan on S&P 500 Index (in Years)

Year (each data point represents a rolling seven-year average of average lifespan)

DATA: INNOSIGHT/RICHARD N. FOSTER/STANDARD & POOR'S
*Note: This graph combines the data from two separate reports.

small business, it doesn't matter. Our days are numbered. Organizations are under siege. If we can't learn to adapt, we may never see another century-old company.

While corporate longevity has plummeted, the average holding time for a stock has gone from eight years to five days. What does someone who holds a stock for five days care about? You guessed it—quick returns. And given that 50–90 percent of all trading is now algorithmic (and often high frequency), returns depend on *volatility*. If stocks are steady, nobody makes any money. So the funds, investment banks, media, and exchanges keep the stocks in motion. Economist Benjamin Graham, the father of value investing, once said, "In the short run, the market is a voting machine but in the long run, it is a weighing machine." It appears that the voting machine is winning.

Corporations want share prices to move too, just exclusively up and to the right. Executives are increasingly compensated in stock (in order to align their interests with those of shareholders), and so they all fall victim to a kind of collective quarterly myopia. They need to move the

needle *now*, and that means revenue or earnings growth (preferably both). But investments in organic growth won't mature fast enough to meet their needs. And this is exacerbated by the fact that companies are getting worse and worse at generating profits using the assets they already have.

Corporate *return on assets*—the ratio of a company's profits to what it owns and owes—is one of the best holistic measures of performance. While *return on equity* or *returns to shareholders* are more common, they are too vulnerable to financial engineering. Assets are harder to game. Unfortunately for U.S. firms, economy-wide ROA has been trending downward for decades. It is now roughly one-quarter of what it was fifty years ago. Even the top quartile of companies have seen ROA decline from 12.9 percent in 1965 to 8.3 percent in 2015. Corporations might be getting bigger, but they're not getting better at generating value from their assets.

So what do they do when investors come calling? They cannibalize themselves. They cut costs in research and development. They centralize

Economy-wide Return on Assets (ROA)

Firm performance, as measured by ROA, continues to decline

SOURCE: Compustat; Deloitte Analysis, Deloitte University

functions. They conduct layoff after layoff. They outsource anything that can be done for less outside the firm, often at the expense of quality or service. They impose strict controls, including desperate measures such as travel and hiring freezes during which none of their employees are trusted to make decisions in the company's best interest.

They do all this to hit their numbers. And they *do*. Miraculously they do. Income goes up. Cash goes up. And generating ROA gets just a little bit harder, because what's left behind is hollowed out. But the investors and analysts got what they asked for. Are they ecstatic? No. Actually, they have a question: how is next quarter looking? So imagine being long on cash and short on ideas. What do you do? That's right, you go shopping, for someone to buy . . . or someone to buy you.

There's just one problem: mergers and acquisitions don't work, or at least not the way we've been led to believe. According to one McKinsey study, almost 70 percent of mergers fail to provide the revenue synergies they promise. A similar study by KPMG found that while 82 percent of M&A participants *believed* their deal was a success a year after closing, only 45 percent had done a formal postdeal review to be sure. When the researchers imposed a strict objective benchmark for success, it turned out 83 percent of the deals had actually failed.

M&A may not always add up, but it does create megafirms with a greater ability to control markets and prices. Just look at the airline industry in the U.S., where a string of mergers and acquisitions has resulted in four dominant carriers that have more than 80 percent of domestic capacity. There are now just four big banks. Five major health insurance companies. And in technology it's even worse. Two mobile platforms. One big search engine. One dominant social network. And one "everything store" that bought a $13.7 billion organic grocery chain with share price gains from the mere announcement of the deal. "Too big to fail" and "too few to choose" are the new normal.

In order to keep these established companies growing in perpetuity,

we need a steady stream of startups for them to acquire someday. Which is why it's distressing to learn that the firm entry rate—the number of startups less than one year old as a share of all firms—was cut in half between 1978 and 2011 in the United States. Self-employment among people aged 15 through 34 has been declining since 1990. Even scarier, the firm exit rate—the number of business deaths as a share of all firms—crossed the entry rate in 2008 for the first time since researchers started tracking this data. While it may feel like we're living in an entrepreneurial boom, the reality is more complex.

As a result of both M&A and entry/exit rates, established firms are becoming a bigger and bigger part of the economy. A recent report from the Brookings Institution showed that the share of firms aged sixteen years or older went from 23 percent in 1992 to 34 percent in 2011. That's roughly a 50 percent increase in the space of two decades. Which would be fine, if we did our best work in large organizations. But that doesn't seem to be the case. Among firms that pursue patents aggressively, small businesses produce sixteen times more patents than their bigger brethren. Innovation, it would seem, is somewhat scale dependent.

Surprisingly, *labor productivity growth*—the increase in the amount of goods and services produced by one hour of work—is also facing headwinds. This metric, perhaps more than any other, has always been an indicator of how fast our quality of life was improving. For much of the twentieth century it was steady. But labor productivity growth over the last decade has been crawling along at the lowest rate since the end of World War II. Occasionally now it even comes to a complete stop. In 2016 it was actually negative. This is so perplexing that economists can't even agree on why it's happening. With all the technological innovation of the last ten years, shouldn't we be getting *vastly* more productive? It's almost as if some other force is counteracting all those gains.

While labor productivity may not be growing as fast as it used to, it is certainly higher today than ever before. Between 1973 and 2016 net

productivity rose by 73.7 percent. What do you think happened to wages over that same time frame? If you guessed that they were stagnant, you're catching on. Wages were up just 12.5 percent over those same forty-three years, meaning that productivity has outpaced pay by 5.9x. Meanwhile, the ratio of pay between the CEO and the average worker went from 22.3:1 to 271:1. Move along, folks, nothing to see here.

It seems likely that our economy is demonstrating what systems designers call *graceful degradation*. It's functional even as parts of the system are failing us. This failure isn't catastrophic. Not yet. But there are signs that the mindsets, structures, and incentives present in our economy are faulty. In their haste to satisfy the market, organizations are becoming bigger, less efficient, and less human-centric. But the market is us. We're doing this to ourselves. And in the midst of that dysfunctional feedback loop, we have an obligation to stop and ask if this is really the best we can do.

EVOLUTIONARY ORGANIZATIONS

The antidote to the diminishing returns of the status quo is to *think differently*. Imagine fourteen thousand healthcare workers managed by a headquarters of fewer than fifty people. How on Earth is this possible? It's actually quite simple. Don't manage them. Jos de Blok was fed up with bureaucracy long before he founded Buurtzorg, a neighborhood nursing company based in the Netherlands. In his country it was common for care providers to pay visits to patients at home, particularly those in need of end-of-life care. But much like the rest of the healthcare industry, these services had become soulless and productized. Personal care was out of the question. Some patients saw as many as 150 different nurses in a year. Nurses were treated as automatons to be deployed, measured, and managed for maximum efficiency. And yet, ironically, costs were up and quality was down. The nurses hated it. Patients did too.

To put things right, de Blok set out to build an organization where nurses could pursue their calling to help others unencumbered. That meant completely reinventing the way care was delivered. At Buurtzorg, each community is served by a self-managed team of ten to twelve nurses who have the freedom to care for their patients as they see fit, with the goal of "helping home-based patients become healthy and autonomous." These teams are responsible for the complete end-to-end delivery of care, and they take that duty seriously. Their very first interaction with a patient is often a conversation over a cup of coffee—to get to know the patient, their needs, their support system—so that they can develop a care plan using everything at their disposal (including family members, friends, and neighbors) to improve the patient's health. A nurse might administer medication one moment and then engage the patient's neighbors about the patient's progress and needs the next. It's a completely empowering experience for the nurses (and the patients), which is why it's no surprise Buurtzorg is routinely named the best place to work in the Netherlands.

How does this approach compare with the traditional, financially motivated model? You might assume it's more expensive, the luxury of a quasi-socialist economy. It's just the opposite. Buurtzorg uses less than 40 percent of doctor-prescribed hours and saves the Dutch social security system hundreds of millions of euros every year. In fact, according to Ernst & Young, delivering care in this new way resulted in a 20–30 percent reduction in per-patient costs. Not only is it financially efficient, but it's structurally efficient as well. The firm generates more than €350 million every year, with just six financial administrators and no CFO. Since delivery of care is owned by decentralized teams, Buurtzorg has been able to avoid the seemingly inevitable bloat of bureaucratic leaders and functions for more than ten years.

Stories like this seem almost unbelievable. Yet Buurtzorg is just one of dozens of similar cases—both past and present—that I discovered

during my research for this book. These *Evolutionary Organizations* have found ways to achieve what conventional wisdom says is impossible. They make better decisions faster. They allocate resources dynamically. They form and disband teams fluidly. They innovate both product and process. They grow without losing the culture they love. They work fewer hours but get more done. They protect the planet but maintain outsized profitability. They create prosperity, not just for their shareholders but for employees, customers, and communities. These organizations have been operating in the shadows for decades. Now a growing dissatisfaction with work has pushed everyone— not just a few iconoclasts—to seek a better way of working.

But curiosity and bravery are two different things. What would possess someone like Jos de Blok to disrupt healthcare (of all industries) with such a radically human approach? He certainly didn't do it because of the numbers. And he didn't do it to keep up with his competition. He did it because, in his view, it was the right thing to do. And that highlights what is profoundly different about Buurtzorg's founder: his mindset. At a very fundamental level, de Blok's mental models of human motivation, how healthcare works (or doesn't), and what success looks like are dramatically different from his contemporaries'. And that alternative worldview allows him to see possibilities others cannot. It allows his organization to do things others can only dream of.

As rare as someone like Jos de Blok might be, he is not alone. Evolutionary Organizations may operate in vastly different categories and cultures, yet what I've found in talking to them is that these communities all share two crucial mindsets. When it comes to human nature, development, and motivation, they are *People Positive*. But they also share a deep appreciation for the dynamic and interconnected context in which we operate, a mindset I refer to as *Complexity*

> We can't change anything until we get some fresh ideas, until we begin to see things differently.
>
> —James Hillman

Conscious. Together these provide the scaffolding for a new way of working based on fundamentally different assumptions about people and the world we inhabit. These mental models, and the dynamic tension between them, are central to the work we need to do. So let's take a moment to understand them better.

PEOPLE POSITIVE

What does it mean to be People Positive? Just ask Jean-François Zobrist, the former CEO of brass auto parts foundry FAVI. When he took the reins of this five-hundred-person firm in 1983, things were pretty bleak. Workers punched in and out on a carefully monitored time clock. Arrive five minutes late and five minutes' salary would be deducted from your pay. In the summer, workers toiled in sweltering heat with the windows closed to earn a "heat premium" for keeping the temperature high. And even getting a new pair of work gloves was a laborious process. Employees had to get a voucher from their boss, walk to the warehouse, wait around for someone to help them, exchange the voucher for what they needed, and head back to work. The cost of their downtime far exceeded the cost of the gloves.

Zobrist believed that if you trust people, they will do the right thing. So he tried something radical. "I came in the day after I became CEO, and gathered the people. I told them, 'Tomorrow when you come to work, you do not work for me or for a boss. You work for your customer. I don't pay you. They do. Every customer has its own factory now. You do what is needed for the customer.'"

FAVI organized itself into a dozen or more "minifactories" that operate as they see fit. No time clocks. No quotas. No management. Just a commitment to doing what is right for the customer and one another. Now a worker at FAVI can buy a whole new piece of equipment without a voucher. And the results are impressive. They've actually managed to

reduce prices by an average of 3 percent per year. And they haven't shipped a late order in more than twenty-five years. Why? Because teams at FAVI have end-to-end control and responsibility for what happens on their watch. On-time delivery isn't a goal; it's a social expectation. Beyond that, workers aren't incentivized by local performance indicators that might pit them against one another. They simply get a share of the total profits of the firm. Because of this, in a good year, workers will make the equivalent of sixteen months of salary. It all adds up. While their competitors have long since outsourced their factories to China, FAVI is the one European holdout in its space, with roughly 50 percent market share. Today FAVI exports *to* China.

What this story illustrates is that our way of working is a reflection of our assumptions and beliefs about human nature. Are we trustworthy or undependable? Are we lazy or industrious? Can we self-manage or must we be controlled? Human behavior, development, and motivation have been subjects of fascination for ages. But in the last century two contradictory schools of thought have collided in the battle for our working lives.

The dominant view, the one held by Legacy Organizations the world over, paints a mechanistic picture of who we are—conditioned by rewards and punishments, our futures dictated by our pasts. This mindset assumes that we don't really want to work or learn without an incentive to do so. Ergo, we must be coerced and controlled in the service of organizations and society.

The minority view, the one held by FAVI, Buurtzorg, and Evolutionary Organizations everywhere, contends that we have an innate desire to fulfill our potential and self-actualize. Rooted in the work of psychologists such as Abraham Maslow and Carl Rogers, this view maintains that people are naturally motivated and capable of self-direction. That we are worthy of trust and respect. And importantly,

that we are all chameleons, capable of suppressing or distorting our true nature depending on our environment and social context. Ultimately, the People Positive leader believes that when it comes to human beings, we show up the way we are expected to show up—the way we are treated.

Had you met Douglas McGregor in his office at MIT in the early 1960s, he might have pulled you into his office to explore this dichotomy, asking which of these two statements best described you:

> There may be said to be two classes of people in the world; those who constantly divide the people of the world into two classes, and those who do not.
>
> —Robert Benchley

X) I dislike work and will avoid it if I can.

Y) I find work to be natural and fulfilling.

If you're like me, you'd probably choose Y. After all, given the right conditions, you do find work meaningful and satisfying.

But then he might ask you which one best describes factory workers, fast food employees, or your colleagues. If you answered truthfully, here you might choose X. I mean, just look at their behavior, right? But not so fast.

In 1960 McGregor published *The Human Side of Enterprise*, in which he introduced two theories about human behavior and motivation that he called Theory X and Theory Y. In his view, each theory was built upon a set of assumptions about human nature, that, if acted upon, would lead to very different management styles.

Theory X reflected the dominant assumptions of his day:

1. The average human being has an inherent dislike of work and will avoid it if he can.

2. Because of this human characteristic of dislike of work, most people must be coerced, controlled, directed, threatened with punishment to get them to put forth adequate effort toward the achievement of organizational objectives.

3. The average human being prefers to be directed, wishes to avoid responsibility, has relatively little ambition, wants security above all.

Considering the modern world of work, the assumptions of Theory X still feel remarkably present. Our job description tells us what to do. Our manager tells us how to do it. Our office hours tell us when to come to work. Our meeting invites tell us which conversations we can participate in. Our performance review tells us how to learn and grow. They all smack of distrust that stems from the simple belief that we're incapable of taking responsibility for our own contribution and development. McGregor saw this belief in the "mediocrity of the masses" as inherently limiting. No matter how well intentioned, Theory X management practices, in his view, would always lead to suboptimal performance.

Theory Y reflected an alternative motivation theory:

1. The expenditure of physical and mental effort in work is as natural as play or rest.

2. External control and the threat of punishment are not the only means for bringing about effort toward organizational objectives. Man will exercise self-direction and self-control in the service of objectives to which he is committed.

3. Commitment to objectives is a function of the rewards [in terms of ego and self-actualization] associated with their achievement.

4. The average human being learns, under proper conditions, not only to accept but to seek responsibility.

5. The capacity to exercise a relatively high degree of imagination,

 ingenuity, and creativity in the solutions of organizational prob-
 lems is widely, not narrowly, distributed in the population.

 6. Under the conditions of modern industrial life, the intellectual
 potentialities of the average human being are only partially
 utilized.

These two lists force us to confront a cognitive bias that we hold about our exceptionality. As individuals, we each identify with Theory Y. *I want to achieve great things. I am creative. I am responsible.* But what about everyone else . . . our colleagues? It's much easier to label them as Theory X. While we have direct access to our own thoughts and feelings, we have no such window into the minds of other people. We have to judge them purely by their behavior, and their behavior is shaped (at least in part) by their environment. If you place someone in a workplace that pays them by the hour, scrutinizes their every move, and treats them like they are disposable, they may (gasp) *present* with Theory X behavior. So we live with the cognitive dissonance that we are (somehow) capable and others are not.

Pay attention to the conversations happening in your office. What kind of language is used to describe people and teams? If you hear words such as "incentivize . . . control . . . enforce . . ." and phrases such as "they can't handle . . . they don't need to know . . . they don't get it . . . ," chances are Theory X is driving a lot of the decision making in your organization, albeit unintentionally. Notice this and call it out. Ask yourself and your peers, "Does how we talk reflect what we actually believe about people?"

McGregor's message may have been ahead of his time, but his work found new life and empirical support in the decades that followed. In the early 1970s, research psychologist Edward Deci began to turn the field of human motivation on its head. He asked students in one early experiment to solve a puzzle cube. One group was paid for every solution they discovered; another was not. When participants were offered a break

and left alone, the paid students kept working at the puzzle longer than the unpaid group. But when the reward was later removed, these students were less motivated to work on the puzzle during their breaks, even compared to the students who had been unpaid the whole time. Something about the reward had *disrupted* their intrinsic motivation. Yet in a variation of the study, *verbal praise* in place of a financial reward actually had the opposite effect, increasing motivation throughout.

Deci went on to create more experiments in this area, culminating in a book called *Intrinsic Motivation*. This controversial publication caught the eye of graduate student Richard Ryan, who invited the author to lunch. A powerful partnership was born. Over many years of collaboration and study, their groundbreaking work evolved into *self-determination theory*. They proposed that we have three innate psychological needs that drive and shape our behavior: autonomy, competence, and relatedness. Perhaps the most important among these is autonomy—the ability to make choices and direct our own lives, free from the control of others. It's so fundamental that Deci and Ryan have routinely referred to intrinsic motivation as "autonomous motivation." Their theory states that we come into the world naturally curious and ready to explore. We don't lack motivation to learn and grow; it's built in. And this innate sense of autonomy can be enhanced or undermined.

In the context of work, the implications are pretty simple. Increase autonomy, and motivation thrives. Decrease it, and motivation erodes. In a study by Cornell University following 323 small businesses, those that offered autonomy grew at four times the rate of top-down, control-oriented firms and had one-third the turnover.

Of course, with great power comes great responsibility. In 1990, on the heels of the Gulf War, orders at FAVI took a nosedive. Zobrist knew there wouldn't be enough work to go around, and the easiest solution would be to lay off the temporary workers who made up 10 percent of the

company's workforce. But he could imagine how his workers might react. So he stood on a stack of empty pallets and addressed the entire factory. He asked what could be done that might preserve the commitments they'd made to these temporary workers and one another. After a hearty session of Q&A, an operator named Gerard piped up, "And if we stopped working for one week [per month] without being paid, could we keep temporary workers longer?" Indeed, Zobrist confirmed, that would help. He called for a show of hands. The vote was unanimous—what amounted to a 25 percent pay cut to preserve the integrity of the community until the lull in orders subsided. Shortly thereafter, it did.

We cannot do the best work of our lives under the auspices of an OS that presumes our stupidity, our laziness, and our untrustworthiness. When it comes to people, in many ways you get what you design for. Evolutionary Organizations know that if you treat people like mercenaries, they will become mercenaries. Treat them like all-stars and they will become all-stars. To be People Positive is to assume and expect the best of everyone.

COMPLEXITY CONSCIOUS

What does it mean to be Complexity Conscious? If you ask executives why the old ways don't work as well as they used to, you'll hear one response more than any other: complexity. Google searches for the term "VUCA"—an acronym that stands for volatility, uncertainty, complexity, and ambiguity—are up almost 1,000 percent since 2004. These words seem to show up in almost every board meeting, earnings call, and leadership retreat. Somehow, complexity has become a catchall for our problems. Subprime mortgage crisis? Complexity. Syria? Complexity. Facebook's role in the U.S. election? It's complex. Suddenly the world is full of surprises.

We know what complexity *feels* like. But most of us don't know what it actually *means*. We use words such as "complicated" and "complex" interchangeably, as synonyms for anything we find confounding.

Picture the **engine** inside a car. Is it complicated or complex? Make up your mind, and hold that thought.

When I challenge audiences with this question, roughly half the people confidently vote for complicated, while the other half vote for complex.

Now, what about **traffic**? Complicated or complex?

Again, I ask for a show of hands. Now, more sheepishly, about a third of them raise their hands for complicated. Another third vote for complex. The remainder suspect my trickery and keep their hands down. Sitting with their confusion, everyone starts to realize the point: we don't know what these words mean.

Contrary to popular opinion, among people who study systems theory, "complicated" and "complex" are distinct words with precise meanings. The engine inside a car is *complicated*. A complicated system is a causal system—meaning it is subject to cause and effect. Although it may have many parts, they will interact with one another in highly predictable ways. Problems with complicated systems have solutions. This means that, within reason, a complicated system can be fixed with a high degree of confidence. It can be controlled.

This is not to say that a complicated system can't be confusing or inaccessible to the layperson. Quite the contrary. Understanding a complicated system, such as an engine or a 3-D printer, requires specialized expertise and experience. Here, experts can detect patterns and provide solutions based on established good practice. This is the domain of the

mechanic, the watchmaker, the air traffic controller, the architect, and the engineer.

Traffic, on the other hand, is complex. A complex system is not causal, it's dispositional. We can make informed guesses about what it is *likely* to do (its disposition), but we can't be sure. We can make predictions about the weather, but we cannot control it. Unlike complicated problems, complex problems cannot be *solved*, only *managed*. They cannot be *controlled*, only *nudged*. This is the domain of the butterfly effect, where a small change can lead to something big, and a big change might barely make a dent. Here expertise can be a disadvantage if it becomes dogma or blinds us to the inherent uncertainty present in our situation.

Complex systems are typically made up of a large number of interacting components—people, ants, brain cells, startups—that together exhibit adaptive or emergent behavior without requiring a leader or central control. As a result, complex systems are more about the *relationships* and *interactions* among their components than about the components themselves. And these interactions give rise to unpredictable behavior. If a system surprises you, or has the potential to surprise you, it is likely complex. Software is complicated. Creating a software startup is complex. An airplane is complicated. What happens between the people on board is complex. An assault rifle is complicated. Gun control is complex. Building a skyscraper is complicated. Cities are complex.

And what about the organization itself? When ten or ten thousand people come together around a mission, what is the nature of that collective? Is it a *complicated* system that an expert can fix or change at will? Or is it a *complex* system—full of uncertainty and surprises—that we must interact with to understand and shape? The answer is intuitively obvious. While many of the activities and outputs of organizations are indeed complicated, *the organization itself is complex.* Accordingly, organizational culture isn't a problem to be solved; it's an emergent phenomenon that we have to cultivate.

It is entirely possible that you get unpredictable behavior out of predictable rules.

—Frank Heppner

The mainstream view is that performance is the result of compliance. *If we can just get everyone to do exactly as we say, we will achieve our goals.* This translates into a culture buried in *governing constraints*—rules, policies, or processes for every imaginable scenario that dictate exactly what should be done. But Complexity Conscious leaders view performance as the result of collective intelligence, emergence, and self-regulation. *If we can just create the right conditions, everyone will continually find ways to achieve our goals.* This translates into a culture that is made coherent and free by *enabling constraints*—agreements that create freedom to use judgment and interaction in the vast majority of situations.

We fail to recognize this on a massive scale. How many reorganizations and restructurings are rolled out with the sentiment "This time we've got it right"? How many leadership training courses have been steamrolled over armies of managers? How many companies have tried to change their cultural values with posters and polo shirts? Every five-year plan, every annual budget, and every fixed target is a public confession that we don't understand the nature of our organizations. Our desire for control blinds us to the truth.

Say you're the CEO of a company where travel spending has spiraled out of control. Your CFO is pushing for a solution. What do you do?

You could approach this issue like a *signal-controlled intersection*. Institute a travel freeze. Create a policy that sets a spending cap on short, medium, and long-haul travel. Require written manager approval on all travel before it's booked. Partner up with a booking engine that can enforce this policy and offer only preferred airlines and hotels. You could set up an audit committee to flag violations and penalize those employees. That *could* work.

Or you could try to create a solution that is a bit more like a *round-about*. Share total travel spend by team (or by individual) with everyone in the company transparently, so they can see how they compare. Share industry averages and historical averages for your firm. Ask for everyone's help in optimizing spend, illustrating how it will drive overall profitability (and thus profit sharing, if you offer it). Start a practice of frequent travelers sharing travel hacks and pro tips at your all-hands meeting. Model what good choices look like with your own behavior. Then stand back and see what unfolds.

To the legacy leader, everything still looks like a factory. And all our problems can be fixed if we work long and hard enough. But our bureaucracies are no match for complexity. They can't handle the surprises that we face every day, and worse, they'll never *surprise us* with an unexpected breakthrough. If we continue to treat the complex like it's complicated, we'll spend our careers frustrated that control is always just beyond our grasp. Gripping so tightly, we'll forget about the magic that can happen when we let go.

BEING BRAVE

You may know the story of Dick Fosbury, who at just twenty-one years of age won the Olympic gold medal for the high jump by going over the bar backward and headfirst. His unorthodox technique, known as the "Fosbury Flop," would change the sport forever. No world record holder since 1978 has used anything but the flop.

What you might not remember, though, is how his competition initially responded. By the time of the Olympics, his teammates in Oregon had been attempting Fosbury's move for two years with no success. Even after his victory, the general sentiment was that this bizarre method worked, *but only for Fosbury*. It would *never* work for them. I face this

every day when I share stories of Evolutionary Organizations with leaders struggling to unlock the potential in their teams.

> "That works for Buurtzorg, but it will never work for us. Must be a European thing."

> "Netflix is a technology company. Of course *they* can do that."

> "I don't know how Handelsbanken gets away with that. Our regulators would *never* let us work that way."

Why do we reject innovative approaches when they are clearly better? Behavioral economists have a name for this: the status quo bias. Think of this as our tendency to like things to stay the way they are. For example, imagine you've inherited a large sum of money and have to decide how to invest it. Here's the twist: It's *already* invested in a specific portfolio of stocks and bonds. Will you make a change? Studies show you probably won't. The same is true for employee participation in 401(k) savings plans. The investment company Vanguard found that if these programs were *opt-in* by default, participation after six months on the job was only 34 percent. But if the program was *opt-out* by default—the equivalent of unchecking a box—participation jumped to 90 percent. We may like to think we're making rational choices all the time, but the fact is we're biased. We prefer to stick with what we've got.

It's not surprising, then, that our way of working has a certain inertia to it. The status quo bias adds to the illusion that the current way is the only way. We strive for improvements—better bosses! simpler budgets! fewer layers!—but they're inherently incremental. We tweak the recipe without ever questioning whether we should make an entirely different dish.

But incremental improvement can be a trap. Scientific Management reveals only the best way to do what we're *already doing.* True innova-

tion often requires a departure from the safety of the status quo. Astro Teller, captain of moonshots at Alphabet's X (formerly Google X), puts it this way: "It's often easier to make something 10 times better than it is to make it 10 percent better. . . . Because when you're working to make things 10 percent better, you inevitably focus on the existing tools and assumptions, and on building on top of an existing solution that many people have already spent a lot of time thinking about . . . But when you aim for a 10x gain, you lean instead on bravery and creativity—the kind that, literally and metaphorically, can put a man on the moon." Sometimes it's easier to start with a blank piece of paper and ask, What are we trying to do? But to do that, we have to release our white-knuckle grip on the present.

Which brings us to one of the most important things leaders and teams need to internalize: our way of working is *completely made up*. This isn't the way it has to be, or even the way it always was. Our way of working was created, brick by brick, by gurus, industrialists, robber barons, unions, and universities—generations of managers and workers who came before us. We can thank them for what is still serving us, and we can change the rest.

Statistician George Box once said, "All models are wrong; some models are useful." As we forge ahead, I invite you to discover what is useful *to you*. You don't have to try everything I present to unlock a better way of working. You just have to go far enough to escape the gravity of the status quo. You have to go over backward and headfirst, even if it feels a little uncomfortable. There are ways to get the support you'll need without politics or propaganda. There are ways to start small *and* strong. This is your chance to do the best work of your life. Don't hesitate.

Part Two

THE OPERATING SYSTEM

94% of problems in business are systems-driven and only 6% are people-driven.

—Attributed to W. Edwards Deming

BECOMING A PEOPLE POSITIVE AND COMPLEXITY CONSCIOUS organization can be overwhelming. It's hard to know where to begin, or what matters most. But through the careful collection and tagging of hundreds of unconventional practices from around the world, I found that Evolutionary Organizations are converging on twelve domains as the proving ground for the future of work. It is in these spaces that the courageous few are taking risks. And it is in these spaces that struggling enterprises will likely find their faults. Together they form a canvas—an Operating System Canvas—through which we can see and shift our organizational identity.

The OS Canvas

PURPOSE

How we orient
and steer

AUTHORITY

How we share power
and make decisons

STRUCTURE

How we organize
and team

STRATEGY

How we plan
and prioritize

RESOURCES

How we invest
our time and money

INNOVATION

How we learn
and evolve

WORKFLOW

How we divide
and do the work

MEETINGS

How we convene
and coordinate

INFORMATION

How we share
and use data

MEMBERSHIP

How we define
and cultivate
relationships

MASTERY

How we grow
and mature

COMPENSATION

How we pay
and provide

Each domain of the OS Canvas asks us to consider an aspect of our organization more deeply than we typically would. For example, what *is* authority? How *should* it be distributed? And how *does* that manifest (or not) in your culture? How *do* you make decisions? How *should* you? Is your approach to authority a signal-controlled intersection or a round-about? Is it People Positive and Complexity Conscious? The canvas forces us to confront the deltas between our assumptions, our beliefs, and our reality. If we say we want to hear every voice but spend most of the day talking over others, that tells us something. If we say we value agility, but every decision requires a dozen approvals, the opportunity is clear.

In the pages ahead, we'll explore how each of these domains is changing, the provocateurs that are shaping them, and the emerging principles and practices they're pioneering. Each domain is broken into five parts: an *overview* that introduces the concept, *thought starters* designed to challenge your assumptions, ways to *take action* and try something new, insights on navigating the domain in *change*, and *questions* to consider as you reflect on and reinvent your own OS.

You may have noticed that the domains of the canvas are generic and value agnostic. That's intentional. We want to ensure that any organization can leverage the canvas regardless of its organizational philosophy. The Morning Star Company, for example, has found huge success in the domain of *structure* by revolutionizing traditional job titles and roles. Every year, four hundred full-time employees at the world's largest tomato processor write their own job descriptions. They do this by authoring a Colleague Letter of Understanding, or CLOU, that contains their commitments to and agreements with one another. CLOUs are reviewed and challenged by colleagues who offer *advice*, not mandates, about what should change. Since this document changes every year, there's no need for traditional job titles or promotions. But that's okay, because everyone adjusts their own salary as they learn and grow. The math works

out. While their industry grows around 1 percent a year, Morning Star has averaged double-digit revenue and profit growth for the past twenty years. Today it generates more than $700 million in revenue. In an industry that normally treats workers as expendable, it has managed to create a way of working that rivals any unicorn for innovative and human-centric design principles.

But this approach to structure may not be right for your context and culture. Your approach may be more or less radical or aligned in spirit but different in practice. That's fine. My only ideological prescription is that People Positive and Complexity Conscious mindsets have the power to reshape these spaces for the better. Every culture has elements of the traditional, the contemporary, and the idiosyncratic. The canvas is a tool for reflection and sensemaking, not judgment.

> The test of a first-rate intelligence is the ability to hold two opposed ideas in the mind at the same time, and still retain the ability to function.
>
> —F. Scott Fitzgerald

Further, this canvas is not intended to be mutually exclusive or comprehensively exhaustive. From a complexity perspective, reducing an organization to its independent parts is folly. The canvas simply highlights the areas that our research tells us are most in flux. Better to start in these dynamic spaces than to remain immobilized by the sheer intractable nature of it all.

At some point in this tour of the OS you're going to start to wonder, How the hell do I lead my organization through a change as profound as these cases and stories suggest? And what if it doesn't work? Don't let that slow you down. The remainder of the book is dedicated to sharing all the lessons my colleagues and I have learned in the trenches with organizations trying to make it to the other side of the rainbow. The transition to a better way of working can be made. But not with the change management they teach in business school. You'll need every ounce of your People Positive

and Complexity Conscious conviction, and more than a few of the tips and tricks you're about to discover.

As we dive deep into these twelve domains, just remember: The problem isn't your leaders. It's not your people. It's not your strategy or even your business model. It's your Operating System. Get the OS right and your organization will run itself.

PURPOSE

How we orient and steer;
the reason for being at the
heart of any organization,
team, or individual.

In 1970 Milton Friedman famously said, "The social responsibility of business is to increase its profits." To put it bluntly, the business of business is business. In the decades since, Legacy Organizations have internalized this to an astounding degree. As we've seen, this maxim has led corporations to optimize everything in society—the market, the law, even our attention—in order to drive short-term gain. At the same time, the cost to humanity and the environment has been profound. Unchecked growth has created the conditions for a climate crisis that is unfolding in real time. This singular focus has also led to rampant inequality and a level of worker engagement that is pathetic at best. A mission statement that places shareholder value as the *definition* rather than the *result* of success is uninspiring. Jim Barksdale, former CEO of Netscape, once quipped, "Saying that the purpose of a company is to make money is like saying that your purpose in life is to breathe."

Instead we can elevate purpose above all. Given that we spend so much of our lives at work, wouldn't it be nice if that work were worthwhile? If it delivered meaning and connection? Take Whole Foods, for instance. If you were to read its "Declaration of Interdependence," originally authored in 1985 by sixty team member/volunteers, you'd see that the company's purpose is to "Nourish People and the Planet." Five words, but a lot of information. Now, what about grocery giant Kroger? Why does it exist? Its stated mission is "to be a leader in the distribution and merchandising of food, health, personal care, and related consumable products and services." Yawn. Imagine showing up every day for forty years with *that* as your rallying cry.

Purpose can be socially positive *or* socially destructive. After all, the key difference between a charity and a terrorist organization is intent. Which is why Evolutionary Organizations aspire to *eudaemonic* purpose—missions that enable human flourishing. And what of profit? Profit is the critically important fuel that powers our purpose. It's the air we breathe. Without it we can't scale our impact or realize our vision.

Which is why the vast majority of Evolutionary Organizations are quite profitable. In fact, the socially conscious and purpose-driven companies featured by professor and author Raj Sisodia in *Firms of Endearment* have outperformed the S&P 500 by a staggering 14x over a period of fifteen years, ten of which were after the publication of the book.

A great purpose is aspirational, but it's also a constraint. It focuses our energy and attention. It places a boundary around our efforts by saying, *Here is where we will build our dream*. Too mundane (e.g., shareholder value) and we lack meaning. Too vague (e.g., change the world) and we lack focus. Too concrete (e.g., a computer on every desk) and we can find ourselves rudderless after the moment of victory. Done well, purpose unites us, orients us, and helps us make decisions as we go.

Thought Starters

Fractal Purpose. Every organization has a purpose. But not every organization ensures that its purpose is fractal—that it shows up at every level. While we want to avoid bureaucratic alignment exercises, teams should have a coherent narrative about how their efforts serve the whole, even if they're intentionally pursuing a divergent path. The team's purpose serves the same function as the organizational one. Even individual roles have a purpose that, if properly articulated, eliminates the need for lengthy job descriptions. If someone who onboards employees truly delivers "members who are informed, connected, and ready to contribute," do we really need to specify the how?

Steering Metrics. Legacy Organizations are obsessed with measurement, often using it as a form of control—to find and punish weak performance. But when we obsess over metrics, we fall victim to Goodhart's law, which states that a measure that becomes a target ceases to be a good measure. Why? Because human beings will manipulate the situa-

tion in order to move the numbers. Instead, we should think of metrics as guides for *steering* toward our purpose. If we make an app that has a purpose of helping people lose weight, then average time in app is interesting, but only insofar as playing with the app translates to healthier users. It's also worth pointing out that steering metrics should, in fact, result in *steering*. You're looking for quantitative and qualitative signals that will help you sense and respond. If you aren't making decisions and taking action based on your metrics, you're doing it wrong. At my company we used to track how many followers we had across social media. We got worked up about it. Then one day we asked ourselves, Have we ever made a change based on these metrics? Nope. We stopped tracking them that day. Last time I looked they were still going up.

Proxy for Purpose. Don't confuse your customer with your purpose. Customer obsession has become a popular theme of late, modeled to the extreme by Jeff Bezos and Amazon. And it's needed. To ignore the customer or lose sight of their needs, as many large corporate teams have, is deadly. But simply making the customer our purpose is also dangerous. If we act on customer feedback without judgment, we run the risk of regressing to the mean, to our basest tendencies. Henry Ford's supposed quip "If I had asked people what they wanted, they would have said a faster horse" illustrates the idea that sometimes our purpose is to take humanity to new places, places they can't yet see for themselves. The truth is that our customers are a proxy for our purpose. They're our partners in solving the tension between our virtuous intent and their actual needs. Whole Foods can't tell if they're nourishing people writ large. But they can tell if more people chose to buy organic fruit last week, or purchased items that contained less sugar, or submitted more positive feedback about their experiences in the store. And that information might be enough for the brand to take a few more steps in the right direction.

Purpose in Action

Essential Intent. Purpose statements, even when they're done well, are sometimes hard to translate into the here and now. In Greg McKeown's best-selling *Essentialism*, he put forth the breakthrough notion of an *essential intent*, a goal that sits between your ultimate vision and your quarterly objectives. He says that an essential intent "is both inspirational and concrete, both meaningful and measurable. Done right, an essential intent is one decision that settles one thousand later decisions." Think of essential intent as a stepping-stone. If we achieve it, we move further along our path to purpose. Tesla's mission is "to accelerate the world's transition to sustainable energy," which doesn't really help an engineer make a decision today. If their essential intent is to create the first *affordable* and *desirable* electric car—and ship 500,000 of them before they run out of money—that gives everyone involved a lot to go on. They'll make trade-offs to achieve affordability, but not at the expense of desirability. And they'll focus on production, knowing that delivery is make or break. Of course, time will tell how well they deliver on these ideas.

Ask every team in your organization to articulate their essential intent. What has to happen in the next six to twenty-four months to keep us moving toward the organization's purpose? Share and discuss over drinks one afternoon. Resist the urge to make them all fit together perfectly. Instead, notice and discuss divergence and convergence. Offer everyone the chance to revise and refine their essential intents regularly, and keep them somewhere everyone can access them.

Six Months or Thirty Years. Here's a slight twist on the same theme. In 2012, around the time Facebook reached a billion users, it published a little red book for employees that contained a lot of the stories, princi-

ples, values, and folklore of the business, memorialized for the next generation. Tucked inside was a page that read, "There is no point in having a 5-year plan in this industry. With each step forward, the landscape you're walking on changes. So we have a pretty good idea of where we want to be in six months, and where we want to be in 30 years. And every six months, we take another look at where we want to be in 30 years to plan out the next six months." While Facebook may have changed these time frames over the years, the spirit of the exercise remains. Clarify your purpose so that you can see it three decades down the line. Then tighten up your road map for the next half year.

Purpose in Change

While OS change is anything but linear, we have found that other dimensions are often dependent on a clear and compelling purpose. For example, distributing authority without clarity on what we're trying to accomplish can lead to empowered people launching projects aimlessly. This results in emergence at best and chaos at worst. Don't start that way. Ensure that any group in transformation—whether it be a team, a unit, a function, or the whole organization—has a strong sense of their collective purpose.

Questions on Purpose

The following questions can be applied to the organization as a whole or the teams within it. Use them to provoke a conversation about what is present and what is possible.

> What is our reason for being?
> What will be different if we succeed?
> Whom do we serve? Who is our customer or user?

> What is meaningful about our work?

> What measures will help us steer?

> How does our purpose help us make decisions?

> What are we unwilling to compromise in pursuit of our goals?

> Can our purpose change? If so, how?

What does it mean to be People Positive about purpose?
Recognize that motivation is connected to a sense of purpose,
meaning, and belonging. Take the time to discuss the connection
between each individual's calling and the organization's
collective purpose.

What does it mean to be Complexity Conscious about purpose?
Accept that purpose is recursive—it shapes us, and we shape it.
Purpose enables freedom and autonomy by ensuring coherent
action. And yet it's also an emergent property of the collective.
It may morph and shift as we learn and grow.

AUTHORITY

How we share power and
make decisions; the right
to make decisions and take
action, or to compel others
to do the same.

When you join a Legacy Organization, the default assumption is that you don't have the right to do anything unless you are given permission. This stems from a theory of control that believes the best way to eliminate risk is through compliance. It all starts innocently enough. In the early days, everyone scrambles to get the business off the ground. If they succeed, the established enterprise attempts to protect itself by standardizing approaches, policies, and structures—it discovers and mandates *the one best way*. Over time, red tape builds up until hardly anything interesting can be done without three different approvals. Only very senior managers and people willing to lose their jobs are able to act freely. The rest of the workforce, unauthorized to solve their own problems, develops a sense of apathy and learned helplessness. Engagement dips. Mistakes are made. And more red tape is wrapped around the handle. Stories of mistakes and failures create a culture of fear. The sense at the bottom is that leadership doesn't trust anyone. The sense at the top is that even more compliance is needed—that *everything* must be specified.

Meanwhile, in Seattle, a game developer called Valve produces regular hits and industry-defining products and platforms. The company generates more than a billion in revenue with a head count of around four hundred people, making it more profitable per employee than almost anyone else. Valve accomplishes all this with a highly unconventional approach to authority. It simply lets its employees decide what to work on. No bosses. No reporting. No oversight. Just "vote with your feet" by choosing projects and tasks that you think are worth your time. There are no official job titles or descriptions at Valve, just two high-level expectations that the company outlines in its internet-famous employee handbook. First, it asks everyone to help find more stunning colleagues. Everyone is a recruiter. And second, it points out that every desk has wheels on it. The message is clear: go find something meaningful to do.

Evolutionary Organizations ensure that everyone has the freedom and autonomy to serve the organization's purpose. The default assump-

tion here is that you can do anything, unless a specific policy or agreement prohibits it. We're starting from a position of trust. Rather than centralizing power in a few senior positions, we aim to distribute authority *as much as possible* to teams and individuals at the edge, where the action and the information are. Teams take full responsibility for their work and their way of working. If something is causing tension or stopping them from achieving their purpose, they can attempt to change it, whether it's a line of code or a massive equipment purchase. This is made possible by new forms of decision making that leverage purpose, principles, consent, and advice to ensure that the choices being made are informed and sound. That means taking the time to define the decision rights held by each member, role, and team. *Who* can make a decision and *how* it should be made are critical components of your OS, and can't be left to chance.

What's being asked of our colleagues here is no small feat. The emotional maturity and professional mastery required are significant. But, counterintuitively, more freedom leads to more learning, and more learning leads to better performance. Of course, when everyone is used to taking orders, they may need a nudge. When David Marquet took over the USS *Santa Fe*—a nuclear submarine and the worst-performing ship in its fleet—he vowed to never give another order, only to share his vision for the ship. Every request for orders from the sailors under his command was met with "Well, what do *you* intend to do?" And at first they were unsure. No one had ever asked. But in time, they came prepared. "Captain, I intend to submerge the ship." Marquet would simply reply, "Very well." He primed his people to think for themselves and claim the freedom and responsibility to run the ship. In the years that followed, the USS *Santa Fe* went from worst to first, setting the bar for retention, operations, and promotions.

How many people in how many roles lack the authority they need to do their jobs well? In a dynamic and fast-changing world, preventing

people from using judgment and making decisions is just too slow. A client recently told me the story of an initiative that required *sixteen individual sign-offs* before proceeding. This is our addiction to control masquerading as "risk management." What firms like this fail to recognize is that the *true* risk they face is that bureaucratic immobilization will make them irrelevant.

Thought Starters

Freedom to Fail. The vast majority of leaders I meet believe that their job is to *ensure perfect execution*. Mistakes are verboten. Failure is unacceptable. And so we use our considerable experience (and energy) to run our teams like well-oiled machines. There's only one problem: they are not machines. They are complex, adaptive systems—human systems—in an age of rapid change. By focusing on execution, we limit the potential for growth in the system. By making ourselves indispensable, we make our teams and organizations less resilient. In complexity, our job isn't perfection, it's building a culture that is always learning. And that requires letting go. Of course, clients often respond that they can't trust their teams to make decisions and take action. I remind them that one of two things is true: either they're wrong and their teams are capable of far more than they realize, or they're right and they need to make an immediate change. What is not okay is to linger in a state of distrust and second-guessing. Have the courage of your convictions, one way or the other.

Minimum Viable Policy. Maximizing freedom means minimizing policy. In their book *Rework*, Jason Fried and David Heinemeier Hansson of Basecamp warn that isolated incidents can too easily lead to bureaucracy. "Policies are organizational scar tissue. They are codified overreactions to situations that are unlikely to happen again. They are collective

punishment for the misdeeds of an individual." This is governing for the exception rather than the rule. At Basecamp, they institute a new policy only when something negative happens *over and over* again. Ask yourself, What's the smallest amount of policy required to protect us while preserving the flexibility to learn and act with judgment?

Decision Discipline. There is a surprising inconsistency in Legacy Organizations: while they are extremely restrictive about *who* can make decisions, they are positively lackadaisical about *how* those decisions are made. Behavioral economists, psychologists, game theorists, and others have learned a lot about how to make better decisions. We have the opportunity to internalize a more disciplined approach to decision making based on all this insight. This applies to how we define decision spaces, how we size decisions, whom we involve, how we decide, and how we evaluate the choices we've made. Take the time to find out more about decision science, so that when faced with a big decision you have a whole tool kit at your disposal.

Authority in Action

The Waterline. One organization that has famously challenged traditional authority structures is W. L. Gore, the maker of Gore-Tex (among many other technologies). Its guiding principles include **Commitment** ("We are not assigned tasks; rather, we each make our own commitments and keep them") and **Waterline** ("Everyone at Gore consults with other knowledgeable Associates before taking actions that might be 'below the waterline,' causing serious damage to the enterprise"). While both principles speak to the company's philosophy, the waterline is a particularly powerful metaphor. On a boat, a hole in the hull above the waterline is survivable. The boat can keep sailing. A hole "below the waterline," however, will probably sink the boat. This kind of mistake is

hard to recover from. So Gore associates use the *advice process* outlined below when they're faced with these tough decisions. You can institute a waterline in your firm. What sorts of decisions can everyone make without consulting others? How much authority are you willing to distribute with no strings attached? Make this explicit and push back when colleagues come to you for permission on something above the waterline. Ask them what they intend to do.

The Advice Process. What about decisions that are clearly *below* your waterline? The ones that are really risky? One approach is to grant those decision rights to specific people or roles. But it's often more powerful to leverage the advice process. In many Evolutionary Organizations, anyone can make a big decision, but first they must seek advice from colleagues who have experience with or will be affected by their choice. That means they seek them out, sit down with them, and talk through all the implications of the decision. This takes time and discipline. For an advice process to work, the person making the decision must be humble enough to seek advice and really listen. And advisers must be comfortable with the idea that their counsel may or may not be heeded. This is not *stakeholder management* where dozens of leaders get to tweak the proposal until it's truly average. This is advice, plain and simple. Once a decision has been made, it should be shared transparently along with the rationale and perspectives that shaped it, so that others can learn too. When it works, this process creates a sense of ownership and responsibility that is hard to replicate.

The Consent Process. Sociocracy is a form of governance that came into contemporary practice in the Netherlands more than fifty years ago. Gerard Endenburg, who brought this philosophy into the workplace, had the realization that *consensus* is not only impossible at scale but doesn't reflect how adaptive systems really work. Instead he proposed that *consent* should be the principle that governs organizations and communities.

His argument was that all policy decisions—agreements, rules, roles, structures, and resources—should be made through the informed consent of those impacted by them. Interestingly, in Sociocracy members may consent to using *other* methods of decision making, including distributing authority to specific roles, teams, or elected representatives, provided they consent to the method itself. The process he developed has been refined over the years by practitioners of Sociocracy and was reintroduced in a more structured form by the creators of Holacracy as Integrative Decision Making or IDM. IDM contains a series of "rounds" that act as an algorithm for processing a proposal. Above all, the method prioritizes inclusion and forward momentum. The final round is the breakthrough: someone who objects to a proposal cannot simply veto it but must try to shape it further to make it *safe to try.* This process can be slow at first as we break some bad habits, but in time, teams are able to fly through an agenda packed with meaty decisions. Keep in mind, the creators of this process recommend using a facilitator to keep the group on track. Give the honor to someone on your team who can reliably corral the group. If you want to give the consent process a try, check out murmur.com, a tool we built to enable collaborative decision making without a meeting.

Integrative Decision Making Process

1. **Propose.** Invite a team member to describe a "tension" they're trying to solve, followed by a proposal or recommendation. Proposer speaks, everyone listens.

2. **Clarify.** Going around the table, give each participant the chance to ask questions of the proposer to understand the proposal. Participant asks, proposer answers, everyone listens.

3. **React.** Going around the table, give each participant the chance to react and/or make suggestions that might improve the proposal. This is their one chance to say whatever they want to say. Proposer accepts feedback but does not reply until the next round.

4. **Adjust.** Based on the questions and reactions, the proposer may edit the proposal (or not) and clarify anything they felt was unclear. The proposer may also remove the proposal at this time if they no longer wish to pursue it based on what they've learned.

5. **Consent.** Going around the table, give each participant the chance to voice an objection, if they have one. An objection is defined as "a reason this would be unsafe to try or would cause irreversible harm to the team or organization." This is a high bar by design, as the goal here is progress—a safe step toward resolving the original tension.

6. **Integrate.** Ask objector(s) to work with the proposer to edit the proposal, making it smaller, faster, better, cheaper—whatever is required to achieve consent from both parties. Once all objections have been addressed, the proposal is accepted. Congratulations, you've made a decision together!

Decision Stack. Everything above, from the waterline to the consent process, is about defining the decision rights in your organization. In combination, I often refer to these approaches as a decision stack. Here's a simple process for creating (or refining) your decision stack collaboratively. Gather everyone on your team together and dive in.

1. **Roles.** Ask: What are the roles that we each hold? Try to think beyond job titles and go one layer deeper if you can. Create a roster of roles with basic descriptions in a shared space.

2. **Waterline.** Ask: What decisions can any of us make without advice or permission? Generate as many answers as you can and then discuss them as a group. What do they have in common? What role do monetary risk and social risk play in defining what is safe? Record your observations and takeaways.

3. **Advice Process.** Ask: When should we seek advice? Using the

overview above, define an advice process that will work in your culture and context. Specify expectations for advice seekers and advice givers. Generate a list of the kinds of decisions you'd expect to see engaging the advice process.

4. **Consent Process.** Ask: What decisions should be made only by consent? Identify decision spaces where the integration of multiple perspectives is crucial. Major changes to policy and structure are good places to start. Clarify whether you intend to use IDM or another consent process.

5. **Rights by Role.** Ask: What decisions should be made exclusively by one of our roles? Identify decisions that should be entrusted to specific roles with expertise, experience, or other qualifications. Take the time to map these decision rights to your roles in a format that will be accessible and editable going forward.

6. **Optimize.** Ask: Can we make this simpler? Reflect on your freshly minted decision stack and look for opportunities to increase the number and kinds of decisions that fit above your waterline or in your advice process. Finally, take a moment to gather consent for the whole proposal. When everyone agrees its safe to try, you're ready to go.

Authority in Change

The problem in most cultures isn't a lack of evolutionary principles and practices. Those bright spots are there hiding in plain sight. The problem is that they don't spread. And they don't spread because the vast majority of teams don't feel they have ownership over their way of working. From the roles they play to the meetings they attend to the tools they use, everything has been dictated to them for far too long. Before you run headlong into changing the big stuff, offer your teams the authority to change a few things about how they work together. It'll go a long way.

Questions on Authority

The following questions can be applied to the organization as a whole or the teams within it. Use them to provoke a conversation about what is present and what is possible.

> How is authority distributed?
> Who can tell others what to do?
> What kinds of decisions do we make?
> How do we make important decisions?
> How do we approach risk?
> What is safe to try? What is not?
> What decision rights do all members have?
> What decision rights are reserved for specific roles or teams?

What does it mean to be People Positive about authority?
Recognize that freedom and autonomy feed motivation. Create an environment where it is safe to try and safe to fail, and teams will learn and grow in extraordinary ways.

What does it mean to be Complexity Conscious about authority?
Accept that we are operating in a complex and rapidly changing world where centralized control is simply too slow and disconnected from reality. Push authority to the edge of the organization— where the information is—so teams can adapt and steer continuously.

STRUCTURE

How we organize and team;
the anatomy of the
organization; formal, informal,
and value-creation networks.

A few years ago, Haier CEO Zhang Ruimin was worried that his teams were too slow and far removed from the customer. The internet had transformed their industry, and only companies that were responsive and adaptive to change would survive. So he broke up the company he'd been running since 1984. Except he didn't do it with spin-offs and private equity. Instead, Ruimin introduced a revolutionary new structure called *rendanheyi*, which translates to employees, user value, unity, and system awareness. Describing the model in *strategy+business* magazine, he said, "The enterprise is transformed from a closed system to an open system, a network of self-governing microenterprises with free-flowing communication among them and mutually creative connections with outside contributors. Employees are transformed from executors of top-down directions to self-motivated contributors, in many cases choosing or electing the leaders and members of their teams."

In practice, this meant Haier reorganized its roughly sixty thousand employees into two thousand *zi zhu jing ying ti*, or independent operating units. These ZZJYTs have full P&L responsibility and all the talent and skills they need to create and deliver products to customers. Leaders are elected, and team members move fluidly across teams as the business and their skills evolve. Not only has Ruimin succeeded in transforming his own organization, but when Haier acquired GE Appliances in 2016, the *rendanheyi* model was applied there as well. While there was some initial resistance, the results speak for themselves. In 2017, after a decade of decline, revenue and profits at GEA increased by 6 percent and 20 percent respectively. A sense of potential is coming back to life. Today Haier is the world's largest and fastest-growing home appliance company, leveraging its unconventional structure to thrive amid uncertainty.

In the early days of a startup, there's only one team and it's all hands on deck. While we may have different skills and experience, we come together in a fluid way, jumping in and out of tasks as needed. But with

success, the workload becomes too great, and we have to bring on more people. Suddenly it's harder to keep tabs on who is doing what, and the work gets more specialized. So we start defining roles. You are the CFO. I am the CMO. And so on. In time, those roles grow to become functions, and those functions become silos. We have to work *together* to actually accomplish anything, so projects often require a marathon of communication and coordination. When that alignment proves too challenging, we default to a strong chain of command, losing some of our entrepreneurial spirit to make our structure work better. Now everyone reports to someone who reports to someone who reports to someone.

Chasing scale, we expand globally, launching new teams in new countries. But over time we notice duplicative teams cropping up in multiple places. How wasteful! So we centralize and streamline them under one leader. Now no one can do UX unless they go through Marcy, for example. Sure, it's slower and a bottleneck, but look how efficient we are! We press on. One day we look up and find that we work in a bureaucratic functional matrix, built brick by brick with our own hands. Structural mediocrity has an inertia all its own.

Instead we can aspire for our organizations to act like dynamic networks, not unlike Hollywood. Members of a system that works this way have the freedom to organize (and reorganize) around projects and programs in order to seize opportunity and accelerate their personal growth. They can hold *multiple roles in multiple places* at the same time. They can maintain a more interesting set of relationships across the network—bringing to light the social structure that is present in the shadows of every organization. In place of central command and functions, we see decentralization and federalization. Teams of teams operate as "cells" in the network, each with their own purpose and their own user(s). Talent and information flow freely. Over time the whole thing comes to look more like a marketplace than an org chart. Sure, hierarchies still emerge, but they are hierarchies of influence, reputation, and

the work itself. They are messy. They are dynamic. And they are held lightly.

Thought Starters

Three Structures. Author and self-proclaimed "management exorcist" Niels Pflaeging has a transformative view of the inherent physics of organizations. He claims every organization has three structures: The first is the formal structure, the realm of hierarchy, which exists purely for compliance with the law. The firm must be owned, operated, and accountable, whether it's a corporation or a cooperative. The second is the informal structure, the realm of influence. This is the social network that exists behind the org chart, always present and often leveraged to get things done behind the scenes. And finally, the value-creation structure, the realm of reputation. This is perhaps the least well understood, because most firms confuse their formal structure with the way they create value. When we talk about organizing and teaming in new ways, about the inner workings of Evolutionary Organizations, it is their value-creation structure that we're talking about. All three structures are always present. The only question is how they are developed and emphasized. In the end, structure is about trade-offs. When we conflate the formal structure and the value-creation structure, we end up with an anatomy that maximizes control at the expense of speed, learning, and collaboration.

Centralize vs. Decentralize. One pattern that will be very familiar to denizens of any large organization is the pendulum swing between centralization and decentralization. This plays out in cycles for almost every activity or function in the business. Take sales as an example. Decentralization creates a lot of action at the edge, but also a lot of variation and duplication. Two different regions compete over the same deal. Multiple

versions of sales materials emerge. One hand doesn't know what the other hand is doing. Gasp! So we centralize: one global sales function, all our ducks in a row. But now we're slow. No variation (and thus less innovation). No local flavor or customization. Customers are unhappy. So we decentralize. And on and on. Instead of constantly overreacting, we can use theory to make a more nuanced choice. In a complex world, decentralization offers many benefits, but we need to maintain collective coherence. To achieve this, we can leverage transparency, social pressure, and principles. What simple rules would have to be in place to allow local or decentralized authority and action in perpetuity? And when centralization seems to make sense—as with some software platforms or shared data sources—I use one question to clarify my position: which structure will make us faster and more adaptive? Whatever that is, centralized, decentralized, or somewhere in between, I'm for that one.

Functions vs. Integration. As we saw in the startup example above, it's easy to end up with massive functional silos if you're not careful. Unfortunately, centralized functions (as we currently conceive of them) are not compatible with complexity. They are too slow, too far from reality, and working under misaligned incentives. The alternative is functional integration. Create teams that have as much capability built in as possible. How is Buurtzorg able to support 14,000 nurses with such a small headquarters? Simple: the teams do a lot for themselves. Take recruiting as an example. Who better to recruit and interview a candidate than the team that person will be joining? They have infinitely more information about the skills and style they're looking for. The same is true for sales, service, you name it. Evolutionary Organizations favor units or "cells" of 10 to 150 people who are self-sufficient. As it scaled, W. L. Gore actually practiced a kind of cell *division*. When factories exceeded 150 people, another self-contained factory was built next door. Larger companies often pine for the glory days when they were 80 people and had the

ability to get shit done. What most massive organizations seem to forget is that we can operate as a marketplace of tribes, serving the customer and one another independently and responsively.

One to Many Roles. Here's a fun fact: there's no such thing as a CEO. Sure, many people run around with that title, but it's not really one role; it's many. The CEO is actually one part recruiter, one part spokesperson, one part board whisperer, one part fund-raiser, one part mentor, and so on. Last time I broke it down I counted something like fifteen distinct roles. The same can be said of almost any position. Job titles mask the complexity of the roles we hold, and they limit our ability to shape them and step in and out of them freely. Instead, think of the organization as a rich network of roles that can be filled and shaped by anyone. Don't limit yourself (or anyone else) to one static job description. Recognize that you already hold many roles in many places. Claim them.

Structure in Action

SLAM Teams. One popular way to bust up the functional matrix is by creating a series of mission-based teams. I like to refer to them as SLAM teams. S-L-A-M stands for self-managed, lean, audacious, and multidisciplinary. In practice, this means the team is pursuing an *audacious* goal for the organization with the authority to make decisions and do the work unencumbered. No reviews. No outside leaders disrupting the work. The team is *self-managed*. In addition, SLAM teams are *lean*— small enough to move quickly. But they are also *multidisciplinary*, containing all (or most) of the skills they need to achieve their mission. Introducing a few teams like this is a step toward the responsive structure in place at Haier and other Evolutionary Organizations and never ceases to result in shorter cycle times, higher engagement, and better outcomes.

Dynamic Teaming. Find a part of your business where teams change regularly and turn it into a dynamic two-way marketplace. The rules are simple: Teams have the authority to add or remove members using the consent process. Members have the ability to join or leave teams with reasonable notice. And members can manage their own bandwidth (i.e., hold more than one role). Open roles are shared transparently and members can "apply" by meeting with other members of the team, discussing the work, and sharing what they have to offer. No formal process here, just a search for a two-way fit.

Structure in Change

Because structure is often a proxy for power, managers (particularly middle managers) may be hesitant to do anything that threatens their position. The questions they're asking themselves are "If I'm not a VP, what am I?" and "If I distribute my power, what do I have left?" The breakthrough happens when they realize that they can trade control for participation. Rather than living in one place in the org chart, they can live in many, a mix of roles where their influence is direct rather than indirect. Positional power traded for reputation. By getting closer to the work they can find joy in it again.

Questions on Structure

The following questions can be applied to the organization as a whole or the teams within it. Use them to provoke a conversation about what is present and what is possible.

> How would we describe our current structure?
> How do products, services, geographies, functions, skills, and customer segments show up in our structure?

> What is centralized? What is decentralized?
> What about our structure is fixed? What is fluid?
> What about our current structure is causing tension?
> How would an ideal structure serve us? What benefits would we expect to see?
> Within teams, how do we approach roles and accountabilities?
> How does our structure learn or change over time?

What does it mean to be People Positive about structure?

Recognize that people are capable of self-organizing if the conditions are correct. Create simple rules or agreements about how teams are formed and changed, then let people go where their skills and energy take them.

What does it mean to be Complexity Conscious about structure?

Organize for adaptivity. Ensure that your structure is driven by the periphery—the teams that touch the market—rather than the teams at the center. Allow teams to reorganize continuously rather than in grand gestures every few years. Clarify and specify structure when helpful, but do so in pencil.

STRATEGY

How we plan and prioritize;
the process of identifying critical
factors or challenges and the
means to overcome them.

Ev Williams thinks the internet is broken. Twitter, the site he cofounded, is a clubhouse for the cognoscenti but also teeming with trolls, misogyny, and mindless regurgitation. Facebook is awash in clickbait, memes, and falsehoods. And digital media has become a race to the bottom, where brainless and sensationalist content earns the lion's share of attention. Williams's answer to all this is *Medium*, a social publishing platform he founded to save us from ourselves. When it launched in 2012, he pulled no punches explaining why it mattered. "The current system causes increasing amounts of misinformation . . . and pressure to put out more content more cheaply—depth, originality, or quality be damned. It's unsustainable and unsatisfying for producers and consumers alike. . . . We need a new model."

Medium was certainly that. From the get-go, the platform subverted nearly every convention in online publishing. Rather than a bench of professional writers, anyone could publish. Instead of quantity, it focused on quality. No clickbait, just thoughtful content with titles such as "The Crossroads of Should and Must" and "Design's Lost Generation." Cluttered navigation and loud design were replaced with the cleanest and most elegant reading interface on the web. And in place of a single feed, algorithmic and editorial curation ensured that quality content would find readers who'd appreciate it. Pageviews were out. Total time reading was in. It was as if the *Medium* playbook were to blindly do the opposite of what everyone else did.

And what happened? People showed up. By the end of 2016, the site had sixty million monthly visitors spending 4.5 million hours reading content that spoke to them. But something was brewing behind the scenes. Less than a year later, *Medium* laid off close to a third of its staff. In a short manifesto about the move, Williams explained its rationale. "Upon further reflection, it's clear that the broken system is ad-driven media on the internet. It simply doesn't serve people. In fact, it's not

designed to. The vast majority of articles, videos, and other 'content' we all consume on a daily basis is paid for—directly or indirectly—by corporations who are funding it in order to advance their goals. And it is measured, amplified, and rewarded based on its ability to do that. Period. As a result, we get . . . well, what we get. And it's getting worse." And with that, the strategy at *Medium* now includes one more counterintuitive stake in the ground: eliminating ads and moving to a subscription model for premium content. In response, one headline read "Ev Williams Has Lost His Goddamned Mind." But that's what they always say when they can't see the long game. *Medium*'s new growth target is ten million subscribers. If it succeeds, it'll control a content budget larger than all the magazines you can name combined. Of course, this strategy is starting to sound familiar, isn't it? Could *Medium* be the Netflix of the written word? Can a company that "believes in feeding minds, not mindless feeds" actually succeed? We can hope.

Of course, strategy is usually about market domination, not erudition. It's about *where to play and how to win*. But in an Evolutionary Organization it can be hard to tell the difference between doing something because it's *strategic* and doing something because it's *the right thing to do*. When Starbucks closed for racial bias training, was that *strategic*? What about when outdoor retailer REI decided to close on Black Fridays? Or when Patagonia ran a full-page ad that stated, "Don't buy this jacket," and introduced an initiative to repair, reuse, and recycle garments it had already sold? Without going inside their boardrooms, it's hard to be sure. When purpose is paramount, anything that serves the purpose is strategic, even if it's not overtly commercial. And yet sometimes the market rewards these actions anyway. Welcome to strategy in a complex postcapitalist economy.

Regardless of your values, strategy is about identifying that which is *critical*—the factors that will make the difference—and determining how to leverage what is at your disposal to maximize your chances of

success. Like organizational purpose, strategy is multidimensional and fractal—it's happening on many fronts at many levels. Which means coherence matters. Amazon Prime's free shipping is a brilliant strategy for overcoming the primary friction in online shopping (shipping fees), but it is codependent on other strategies that prioritize market share, cart size, and logistical efficiency. A random collection of trade-offs that would be shrewd in isolation might be disastrous in combination. Which is why the dialogue around strategy is as important as the output. When we maximize cognitive diversity and create space for ideas to collide, we're far more likely to make progressive *and* coherent choices that serve our purpose, whatever it might be.

Legacy Organizations fall down on strategy in two ways. They hesitate to make the kind of bold trade-offs that cultures such as *Medium*, Tesla, or Apple do every day. In their desire to be "the leader," they try to appeal to everyone, do everything, be everywhere, and offend no one. But good strategy is about identifying critical (and often controversial) factors that are going to define the future of a category. Having a pipeline of ever-more-sophisticated BlackBerry mobile devices is great, but it doesn't matter if the world is moving to touch screens. While we often think of this as the job of the CEO, these counterintuitive insights— unpopular by definition—are often hiding in the organization and waiting to be discovered. What's more, many organizations forget how potent the connection between purpose and strategy really is. If you don't have a compelling vision—a dent in the universe beyond shareholder value— your strategies will fall flat. Because how can we win if we don't know what winning looks like?

Thought Starters

Wild Swings and Sure Things. His politics and cantankerous demeanor aside, Nassim Nicholas Taleb introduced an extremely valuable concept

in his book *The Black Swan* called the barbell strategy. This financial strategy is named for how it distributes risk—to two extremes: invest 85–90 percent of your assets in extremely safe instruments, and place all of the remainder in highly speculative bets. The reasoning holds that the "sure things" will create a floor for performance, limiting your downside risk, while the "wild swings" will expose you to potentially life-changing gains. The same strategy can be applied in an organizational context. Your project portfolio is probably a bell curve, but it could be a barbell.

Careful with OKRs. The OKR, which stands for objectives and key results, is a concept introduced to Google and several other prominent firms in Silicon Valley by venture capitalist John Doerr. The basic idea is that each person in the organization should identify their strategic objectives for the quarter and break those down into the more measurable key results that will indicate if they've been successful. It's goal setting with a modern twist. OKRs should be stretch goals, not easily accomplished (to prevent sandbagging), and transparent (to encourage collaboration and understanding). There are two things to watch out for here. The first is, again, Goodhart's law. Once people set their OKRs, they're going to do everything they can to hit them, including things that aren't good for the business. As W. Edwards Deming observed, "People with targets and jobs dependent upon meeting them will probably meet the targets, even if they have to destroy the enterprise to do it." The second is that many firms try to use OKRs as a form of top-down control, ensuring that each subordinate's OKRs fit with the OKRs of their superior. While this feels like alignment, it all but eliminates any chance of divergence or serendipity. A healthy system is not going to cascade in a perfect hierarchy of intent. We want outliers. We want wild swings. Not a lot, but enough. And it's easy to forget to make space for them in a system like this.

Learn Faster. In a truly dynamic marketplace, all winning strategies will have one thing in common: the desire to be faster. Or more specifically, to *learn* faster. In the 1950s, aviator and military strategist John Boyd created the OODA loop—observe, orient, decide, act—to explain how expert fighter pilots continually processed information in the heat of battle. In talking about the OODA loop, Harry Hillaker (chief designer of the F-16) said, "Time is the dominant parameter. The pilot who goes through the OODA cycle in the shortest time prevails because his opponent is caught responding to situations that have already changed." Back in the present, technological monopolies are putting this into practice, to lethal effect. Amazon now averages more than one software deployment every second. And on Reid Hoffman's *Masters of Scale* podcast, Mark Zuckerberg shared how Facebook gets smarter every day. "At any given point in time, there isn't just one version of Facebook running, there are probably 10,000. Any engineer at the company can basically decide that they want to test something. There are some rules on sensitive things, but [. . .] they can launch a version of Facebook not to the whole community, but maybe to 10,000 people or 50,000 people—whatever is necessary to get a good test of an experience." I may not agree with what they test, but the speed and scale with which they do it is impressive. Do you have a thousand versions of your business running? The future belongs to learners.

Strategy in Action

Even Over Statements. The problem with priorities is that we have too many. The word "priority" didn't even have a plural form until the 1900s; before that it meant "the very first thing." Unfortunately, when you prioritize everything, you prioritize nothing. Just because you want absolute low cost and absolute quality doesn't mean you can have both in

equal measure. When push comes to shove, something's gotta give. The "Agile Manifesto" used *"over" statements* such as "responding to change *over* following a plan" to make trade-offs explicit in service of better software. More recently I was introduced to *"even over" statements* by Tom Thomison, one of the cofounders of HolacracyOne. The even over statement, which is a form of strategy, does a great job of indicating that even though we value the thing on the right, we value the thing on the left *more*. Amazon, for example, might use the statement "market share even over margin" to clarify that although profitability is important, at this moment the company would rather invest cash flow in further cementing its dominance in the market. Good strategy statements work in *both* directions—one good thing over another good thing. My former colleague Jordan Husney illustrated this nicely with a set of competing statements about his burgeoning software business. One of the two versions below is real, but the important thing is that they're both plausible. That's how he can be sure the trade-offs are real.

> *Small organizations even over global enterprises*
> *User growth even over revenue conversion*
> *Initial sign-up retention even over existing user retention*
> *Desktop experience even over mobile experience*

Or:

> *Global enterprises even over small organizations*
> *Revenue conversion even over user growth*
> *Existing user retention even over initial sign-up retention*
> *Mobile experience even over desktop experience*

Every ninety days or so, gather your team and try to generate a few strategy statements of your own (or review and update the ones you

have). Think about what you need to prioritize in order to succeed. What important trade-offs don't come naturally? What has been implicit that you can make explicit? Ideally, this happens at every level: the organization, the team, and even the individual. And remember, bad strategy statements contain trade-offs that don't hurt, or ignore them altogether. Good ones make tough choices. Great ones? They make trade-offs that everyone else thinks are crazy, and in the end turn out to be right.

Scenario Planning. Your purpose is clear. Your essential intent is well defined and regularly tuned. Using these beacons, people are "voting with their feet" and energizing the projects that matter most. Your *even over statements* are shaping a thousand smaller decisions. It's fair to ask, Do we even need a plan? Yes, but not a plan that says what *must* happen, a plan that says what *might* happen. Scenario planning has the potential to open us up to possibilities that we might otherwise not consider. Done right, it's an inexpensive way to wrestle with uncertainty. To get started, gather a group with as much diversity as you can fit in the room, and ask them to generate as many possible futures or micro scenarios as possible. Cluster them and discuss the factors involved. Discuss what could be done to mitigate or navigate these scenarios. Listen closely for anything that's too quickly dismissed—the black swans that are overlooked until it's too late. And remember, we're not trying to *anticipate* the future; that's not Complexity Conscious. We're trying to create awareness. Readiness. Preparedness. So that when something unexpected happens—and it will—your team is less likely to be surprised. Your OODA loop is short. And you're ready to act.

Red Team. Over time, it can be hard to separate strategy from the status quo. Are we operating in sixty countries because that's critical to

our success or because we happened to acquire a bunch of foreign offices? The longer we operate, the more organizational debt we likely carry with us, limiting our ability to innovate. Indeed, Professor Clayton Christensen coined the term "the innovator's dilemma" to describe why incumbents are so often disrupted by upstarts and unexpected competitors. To resist this pattern, try launching a *red team*. Originating in the military and intelligence communities but now practiced in a variety of business contexts, a red team is charged with one mission: putting you out of business. Gather any group of colleagues and ask them to design the competitor that would bury you. You'll be amazed how quickly (and how excitedly) they jump to the task. Their enthusiasm is proportional to the amount of org debt you're carrying. With your red team's design in hand, you now have two options. You can go put their best ideas into practice by changing your OS. Or you can turn that red team into Red Team, Inc. and make a real go of it. The choice is yours.

Strategy in Change

One of the processes that stands in the way of reinventing our way of working is the annual planning process. In traditional organizations, this document lays out where we will play and how we will win, and contains implications for structure, budgets, projects . . . everything. The problem is that we put so much effort into creating it that we treat it as gospel. It becomes true north for our activity rather than what we're seeing and learning in the real world. This prevents teams from evolving their OS. Instead, demote your annual operating plan. Make it an annual operating *prediction*. Let teams take ownership of their local expression of strategy and operations. This will leave space for emergence and harness the full potential of your membership.

Questions on Strategy

The following questions can be applied to the organization as a whole or the teams within it. Use them to provoke a conversation about what is present and what is possible.

> What is our current strategy?
> How is our strategy informed by our purpose?
> What are the critical factors that will mean the difference between success and failure?
> What are the trade-offs we're willing to make?
> How do we develop, refine, and refresh our strategy?
> How do we communicate our strategy?
> How do we use strategy to filter and steer day-to-day?
> How does our strategy inform our planning process?

What does it mean to be People Positive about strategy?
Recognize that sound strategy depends on our ability to perceive what's really going on. Each of us has a piece of that truth. Create shared consciousness through forums and tools that integrate different perspectives, and question the logic of your strategy regularly.

What does it mean to be Complexity Conscious about strategy?
Accept that in areas of rapid change, your strategy is only as good as your ability to learn and adjust course. Because complexity is inherently surprising, know that effective strategies may appear unlikely or counterintuitive, until they're suddenly obvious.

RESOURCES

How we invest our time and money; the allocation of capital, effort, space, and other assets.

In Sweden, a century-old financial institution—Svenska Handelsbanken—organizes around the simple notion that, "the branch is the bank." That means more than eight hundred branches are free to make their own decisions about lending, customers, service, marketing, you name it. The head office in Stockholm exists to support and enable these branches. That means no fixed targets from corporate, only relative targets comparing overall performance to the market average and branches to one another. And here's the kicker: this bank abandoned budgeting more than fifty years ago. Jan Wallander, the man who decentralized operations as CEO in the 1970s, had seen enough forecasts gone awry by the time he took over. He was convinced that "either a budget will prove roughly right and then it will be trite, or it will be disastrously wrong and in that case it will be dangerous." Instead the bank relies on dynamic funding and just-in-time spending that is supported by ongoing dialogue among branches and functions. Incentives and fat bonus programs are nowhere to be seen, instead replaced by the Oktogonen Foundation, a truly innovative profit-sharing and stock-ownership program that funnels a share of the firm's profits to employees when the bank outperforms its rivals. More than 10 percent of the total equity is now held by employees. Handelsbanken has been more profitable than the average of its peer banks for more than forty years. It has weathered every major financial crisis in my lifetime with remarkable aplomb, delivering the biggest increase in shareholder value of any European bank over the last ten years.

Every organization, regardless of its beliefs, must decide how it will use the resources at its disposal to pursue its purpose. Every dollar is a chance to invest in *something*—current operations, new initiatives, external investments, distributions to shareholders, share buybacks, or even philanthropic donations. Every hour of human effort is a chance to energize existing roles and projects, embark on something new, strengthen

a relationship, develop a new skill, rest and recuperate, and more. If organizations are like living systems, then resources are oxygen. Every choice we make nourishes or starves this network of possibilities.

The solution to this resource allocation problem, we think, is to create a budget. Perhaps no other process is so respected and reviled as the annual planning and budgeting process. For those who have managed to avoid direct contact with James O. McKinsey's "gift" to humanity, here's how it works:

1. A *business plan* is created through a bottom-up and top-down negotiation between business units and corporate headquarters, in which they make assumptions about what will happen, internally and externally, for the next twelve months.

2. The business "locks" a version of the plan that they believe walks the tightrope between satisfying shareholders and being theoretically achievable. This forms the basis of a *budget* for the year ahead.

3. The budget includes a detailed set of *targets* for sales and profitability, as well as funding *commitments* to support operations and investment. All of this is underpinned by an elaborate set of *incentives* designed to ensure the delivery of the budget as written.

4. During the ensuing twelve months, actual *performance* is compared to the budget with a high degree of scrutiny. Any *variance* is attributed to "good" or "bad" performance by the team. Managers are expected to eliminate any gaps in performance, ideally without reallocating resources. The idea that the underlying assumptions in the budget may have been unrealistic is never seriously considered.

5. Midway through that same year, the process starts all over again for the upcoming year, the resulting workload laid on top of the day-to-day execution of the existing plan and budget.

As I'm sure you're already aware, there are a few problems with this approach. For starters, budgeting is extremely time consuming. Ford Motor Company once estimated that its planning and budgeting process cost around $1.2 billion a year. You read that right: the process by which the company controlled how money was spent cost it *more than the GDP of Grenada*. Many organizations I've worked with are similar, spending half their year winding through the process outlined above at the expense of doing actual work.

Budgets also suffer from being too rigid and unresponsive. In a complex world, circumstances change quickly, but traditional budgets don't. Perhaps there's a downturn in the market as a whole. Or customer preferences change. Maybe a new competitor emerges that disrupts your pricing model, as Warby Parker did to the traditional eyewear brands. An annual budget doesn't allow for that volatility. The interdependence of the plan—the fact that it all has to add up—means that changes, even small ones, create cascades of adjustment that nobody has an appetite for. A leader whose year is going according to plan is actually disincentivized from helping a struggling division change the budget. The system doesn't want to play jazz; it wants to play sheet music.

Most important, budgets can actually work against performance. Dr. Steve Morlidge offers a simple explanation for this in *The Little Book of Beyond Budgeting*. "Budgets also sub-optimize performance by forcing the business to use the same set of numbers for competing purposes. On the one hand, good targets need to include an element of stretch, but forecasts need to be realistic, while cost budgets need to be 'tight' in order to constrain spending. It's simply not possible to fulfill all three purposes using the same number." Because of this, everyone involved is actively trying to bend the process to their advantage. Leaders demand higher targets knowing that actual performance will probably miss, hoping to get what they need. Managers sandbag and propose targets they know they can beat, assuming that they will be forced to aim higher.

Dr. Morlidge continues, "Revenue targets can act as a ceiling since there is no incentive to exceed them, whereas cost budgets become a floor as there is no motivation to spend less."

Budgeting is broken. But few organizations find the courage to bring true dynamism and collective intelligence to resource allocation. Fortunately, it can be done, if we're willing to give up the illusion of control that modern budgeting provides.

Thought Starters

It's All Relative. If a Legacy Organization committed to—and hit—a fixed performance target of 10 percent growth this year, but the market grew 20 percent, does everyone deserve a bonus? The Beyond Budgeting Institute would say no. The institute is an organization and membership network that was founded to solve the issues and frustrations caused by traditional budgeting. Its quest led it to discover organizations who had moved "beyond budgeting" to something more adaptive and human. From these early cases and its own experiences, the institute articulated an alternative approach that is philosophically aligned with, and a source of inspiration for, many of the ideas in this book. The Beyond Budgeting community plays host to a treasure trove of stories and practices that defy convention. Take fixed performance contracts, for example. Many of its member organizations have abandoned *fixed* targets in favor of *relative* ones. Where a Legacy Organization might commit to a specific revenue-growth target for the year, Beyond Budgeting would recommend beating the average performance of their competitors. Why? Because what if the market was down 20 percent and the business stayed flat? That would be a miracle. But under the traditional approach no one would be rewarded. In an uncertain world, relative performance is the only true benchmark. Company to company. Team to team. And before you jump to the conclusion that relative performance might apply

person to person, I'll save you the stress. The Beyond Budgeting folks believe that individual performance doesn't really exist. Performance is a team sport.

The "Tragedy" of the Commons. You may have heard about the *tragedy of the commons*, the idea that when we share a public resource, we ruin it by acting in our own self-interest. Your office refrigerator is a classic example. *Why won't anyone throw away their expired and rotting food! Who is stealing yogurts that don't belong to them?* And so on. The science says we're incapable of managing resources that belong to everyone, right? Not so fast. Elinor Ostrom shared the 2009 Nobel Prize in Economics for her research into how humans successfully (and unsuccessfully) manage the commons. What she found was that successful communities leverage something she calls polycentric governance. In an interview at the Mercatus Center at George Mason University, she explained, "The strength of polycentric governance systems is each of the subunits has considerable autonomy to experiment with diverse rules for a particular type of resource system and with different response capabilities to external shock. In experimenting with rule combinations within the smaller-scale units of a polycentric system, citizens and officials have access to local knowledge, obtain rapid feedback from their own policy changes, and can learn from the experience of other parallel units." By now this should sound familiar (albeit a bit technical). When teams have the right (and the inclination) to experiment with their own rules and norms, they can find ways to collaboratively leverage and preserve resources. If "our money" really were *our money*, who knows how well we could deploy it in service of our collective purpose? The tragedy of the commons is that we think we can't share.

Resources in Action

Zero-Based Zephyr. Zero-based budgeting originated in the 1970s as a way of challenging the inertia of bloated budgets that persist year after year. The idea was to build your budget up from zero, questioning everything as you go. At Morning Star, business units present their plans for the upcoming year and colleagues invest in the ideas they like best using a virtual currency. Inspired by that practice, here's a valuable exercise that will help you see where the wind is blowing:

1. Bring your team or company together in a large room where you have the freedom to move around.
2. Prior to the meeting, collaboratively generate a list of every project, program, and initiative that is currently under way, as well as any fully formed prospective projects that the group would like to see funded and executed in the next quarter. Capture them on large-format note cards or sticky notes and arrange them neatly on the wall or floor.
3. Give each attendee one hundred virtual dollars. You can use Monopoly money, stickers, or even just pens and an honor system.
4. Now ask everyone to invest their money as they see fit. What would they do if they were the sole decision maker for this upcoming period?
5. When all the money is spent, have a discussion about the patterns that have emerged, and allow time for everyone to adjust their investments based on what they've learned.

Now you face a choice: will you act on the wisdom of the crowd?

Participatory Budgeting. In New Zealand a group of individuals and companies focused on social impact have organized as a network known

as Enspiral. They collaborate; hire one another; and share talent, clients, and, most important, ownership over the ecosystem itself. As the network grew, it faced a conundrum about how to invest its own capital. Because the community is both decentralized and diversified, it was not obvious what projects deserved funding or who should decide. Not surprisingly, a group within the community sprang up and invented a whole new tool, called Cobudget, to solve that problem. The application allows a team or community to propose, discuss, and fund new projects in a transparent and intuitive way. This tool is just one part of a broader social trend called *participatory budgeting*. Twenty-two cities have adopted the practice to date, giving citizens the power to direct millions of dollars to more than 1,532 projects that matter to them. To get a sense of how it works, invite your team to try something like Cobudget and put your discretionary budget for the next quarter on the line. You'll see a whole new level of engagement.

Resources in Change

Often there is a desire in teams we coach to reinvent the budgeting process early on. I suspect this is because it is so onerous and time consuming. On the list of what's stopping us from doing the best work of our lives, it ranks *highly*. But my team typically cautions against this. The budgeting and planning process is one of the few practices that truly span the organization and, as such, requires a lot of consent to revolutionize. Even if leadership is fully on board, I'll often start with new ways of working at the team level and come to resources later with more momentum and maturity.

Questions on Resources

The following questions can be applied to the organization as a whole or the teams within it. Use them to provoke a conversation about what is present and what is possible.

> How do we allocate funds, effort, space, and other assets?
> Are resources deployed annually, quarterly, or dynamically?
> Do we use targets, forecasts, trends, and/or tolerances? If so, how?
> How do strategy and planning influence resource allocation?
> How do we balance resources across the short term and long term?
> How do we balance resources across our core business and innovation?
> How do we define and measure the performance of our resources?
> How does our approach enable us to respond to emergent events?

What does it mean to be People Positive about resources?
Recognize that people are not resources, they are people—capable of directing their own time and attention to where they can add the most value. They're also capable of delivering performance without fixed targets or individual incentives. Let relative targets and a share of the wealth created by the business guide behavior.

What does it mean to be Complexity Conscious about resources? Accept that you cannot predict the future. Choosing how and where to spend your money a year in advance is folly. Minimize long-term commitments where appropriate to maximize discretionary funds. Ignore annual rhythms and allocate resources dynamically based on real-time information.

INNOVATION

How we learn and evolve;
the creation of something new;
the evolution of what already
exists.

Growing up, my father would say to me, "Two wrongs don't make a right, but two Wrights make an airplane." Perhaps no story better epitomizes the idea of American innovation than the story of Orville and Wilbur. Over the course of three years and numerous attempts, they invented, built, and flew what is generally credited as the world's first successful airplane. The flight was just fifty-nine seconds and 852 feet, but the aeronautical engineering that went into it—including the invention of three-axis control—is still with us today. Our modern perception of innovation is synonymous with their story. Mad men. Isolated. Focused. Relentlessly pursuing their vision through years of trial and error. Innovation, we believe, must be done by *innovators* in the laboratory. The rest of us should just keep our heads down. Innovation isn't a distributed phenomenon; it's a department.

But innovation doesn't have to be so intentional. It often occurs when things are used in ways we did not intend. Paleontologists Stephen Jay Gould and Elisabeth Vrba coined the term "exaptation" to describe the phenomenon of a biological trait being co-opted for another purpose beyond its initial advantage. Exaptation isn't limited to biology, though; it happens in business too. Corning invented Gorilla Glass in 1960 and shelved it for more than forty years. It wasn't until Steve Jobs came calling that they realized it was perfect for smartphone screens. Alexander Graham Bell invented the precursor to the telephone in order to help deaf students visualize sound. Percy Spencer invented the microwave when a magnetron melted a chocolate bar in his pants pocket. Play-Doh was supposed to be a cleaning product for wallpaper until kids started "misusing" it. While we like to think of modern life as the result of human ingenuity and eureka moments, the truth is that accidental breakthroughs have led to as much success as their more intentional counterparts. Randomness and innovation are good friends.

Of course, this presents a problem. One of the primary goals of global brands and institutions is to *eliminate variation* and *ensure conformity*.

That's the only way to provide a consistent experience. But eliminate variation and you've stopped evolution dead in its tracks. While a centralized innovation hub might be able to produce a few hits, it severely limits the likelihood of exaptation. And importantly, product innovation isn't the only kind of innovation we need. Not even close. Every activity in the organization is a black box of potential energy. If every team isn't constantly learning and improving in ways big and small, we're missing our chance to pursue our purpose with everything we've got. Innovation can happen intentionally or unintentionally, centrally or out at the edge. We need to ensure that we've thought about, and made space for, evolution at all levels.

Thought Starters

Innovation Everywhere. Because of our obsession with efficiency, most organizations like to keep innovation and operations separate. They build things and they run things. But within the last decade, DevOps has emerged as a software engineering culture and practice that has completely upended that notion. As teams release software faster and more frequently, the interaction between development, quality assurance, and operations was stressed. Developers want change. Testers want risk reduction. And operators want stability. The solution was to bring these functions together into one continuous process. When teams are truly empowered, customer focused, and functionally integrated, it can be hard to say where "innovation" happens. Amazon recently confounded the Securities and Exchange Commission when it refused to disclose its research and development spending in the same way other major public companies do. At present, the company discloses only "technology and content" spending of more than $20 billion. When pressed as to why it couldn't break out R&D spend, the firm's worldwide controller wrote, "Our business model encourages the simultaneous research, design,

development, and maintenance of both new and existing products and services. For example, our teams are constantly working to build new Alexa skills and simultaneously maintain current skills, and these activities are within a continuum of those described in ASC 730–10–55–1 and 2 and are not easily distinguishable operationally." The point? We do *so much* innovation *so frequently* in *so many* places here at Amazon that it's not worth tracking project by project the way other companies do.

Watch What Spreads. In his book *Reinventing Organizations*, author Frederic Laloux shares a story of innovation at Buurtzorg that highlights how innovation *moves* when we don't try to force it. A group of nurses within the organization had noticed that when their elderly patients suffered a fall, they often broke their hips, which reduced their autonomy (sometimes permanently). So they created a new program focused on accident prevention and delivered it in their local market. They were so excited with the results that they brought them to Buurtzorg CEO Jos de Blok. This program should be rolled out across the entire company, they said. But instead of assigning a task force or piloting the program in other regions or announcing it as a company-wide initiative, he did something else entirely. He asked the team to write a story about what they'd created and publish it to the company's internal social network, along with a guidebook for how to stand up the program. If the idea was good, he reasoned, it would spread. Before long thousands of Buurtzorg's nurses were working in what are now called Buurtzorg+ teams—delivering home care *and* accident prevention. If teams are truly free to try new things and adopt what works, innovation can become organic. Next time you hear a great idea or see a winning approach, don't roll it out, offer it up.

Biological Barbells. Sometimes it's hard to know how much variation or innovation is necessary. Should you put your resources behind what is

working right now or invest in the future? Nature offers us some clues. By studying complex adaptive systems such as ant colonies, scientists like Deborah Gordon have discovered that ants increase and decrease the randomness of their collective explorations using algorithms that balance information and risk. If a colony is released into an empty room (low/no information) they will fan out in increasingly random patterns. Introduce an apple (valuable information) and they will quickly converge upon it. Interestingly, most *but not all* of the ants will seize on the "sure thing" the fruit represents. Their built-in algorithms ensure that some percentage of their membership continues to look for the next big thing. This harkens back to Taleb's barbell strategy. When we have a sure thing on our hands, we should deploy most but *certainly not all* of our resources there. When we don't know what's going to work—in a startup or new division, for example—we need to increase exploration, randomness, and variation, even if it feels a little counterintuitive.

Innovation in Action

Defaults vs. Standards. It's easy to get in the habit of cementing practices, methods, tools, and products as *standards*. Standards tend to be *enforced*. We use *this* tool and *only* this tool. We evaluate leads using *this* process and *only* this process. And so on. The great thing about standards is that they show us a proven way to do something and they are reliable (for the most part). The problem with standards is that they undermine our ability to use judgment, innovate, and learn. Instead of enforcing standards, think about proven practices as defaults. Defaults are exactly like standards with one exception: you don't have use them. A default says: If you don't know what you're doing, *do this*. If you don't have time to think, *try it our way*. But if you've achieved some level of mastery in an area and you think you see a better way, feel free. Let us all know how it goes, because either you'll generate further proof that our

default is sound or you'll sow the seeds for a new default that we can all benefit from. In the case of my firm, we maintain a default fee structure for our services. If an experienced member wants to try something different—a price premium, a discount, an equity swap—there are no rules against that, but we want to learn from it. Our members know they *can* take a risk and go their own way, but only when they think that we all stand to benefit.

Twenty Percent Time. Google famously offered its early employees the opportunity to spend 20 percent of their time pursuing personal projects that might or might not have a clear connection to their daily work. We have this program to thank for products such as Gmail, Google Maps, Google News, and even AdSense. All of them grew out of projects started in employees' spare time. Valve Software, as we've discussed, went one step further. They offer employees 100 percent time. Work on what compels you; change it up at any time. Now, you may not be ready to go that far, but consider applying Google's model in your team for a ninety-day trial period. Remember, the day-to-day workload is going to undermine this program at every turn. That's what happened at Google in recent years. You'll have to model and support this practice with discipline and passion to realize its many benefits.

The Lean Startup Method. Inventing the future is hard. The early days of any startup (including new ventures inside established companies) are all about the search for product/market fit. The problem is that we fall in love with our initial vision and spend months and years perfecting a product that never sees the light of day. When we finally share it with customers, we're astonished to find that it doesn't really meet their needs or that their needs have changed. The Lean Startup method offers a more scientific approach to new-product development. Developed by Eric Ries, the method proposes a feedback loop containing three stages:

build, measure, and learn. To get started, pick a problem you are interested in and have the agency to solve, and build a minimum viable product (MVP) that will allow you to get user feedback as quickly as possible. You're going to be engaged in what Ries calls *validated learning*. Instead of trusting your assumptions and barreling ahead, you're going to identify them and ask one of the most powerful questions in innovation: how might we validate that?

Innovation in Change

OS transformation rarely starts with a focus on the practice of innovation, because other domains are typically what hold our creative energy back. But in the end, finding our way to an Evolutionary OS is all about change. We want our organizations to be more agile and adaptive. What that means in practice is that we want to get better at learning, better at finding and trying new things. This applies at the product level, of course, where we meet the market, but it's equally important in our way of working. The act of working on any dimension of your OS is an act of disruption and invention. So while you might not get around to talking about the system-wide approach to novelty right away, rest easy knowing that you're working on it.

Questions on Innovation

The following questions can be applied to the organization as a whole or the teams within it. Use them to provoke a conversation about what is present and what is possible.

> What is our philosophy on innovation?
> Where, when, and how does innovation happen?
> Who participates in innovation? Who has the right to innovate?

> How do we approach incremental and disruptive innovation?
> What is the role of failure and learning in innovation?
> How do we evaluate new ideas, approaches, and products?
> How do we balance the short term and the long term?
> How do we manage our portfolio of ideas, prototypes, and products?

What does it mean to be People Positive about innovation?

Recognize that people are inherently creative given the right conditions. Trust them to sense opportunity and pursue it fluidly. A true culture of innovation is one where we can't tell the difference between operations and invention.

What does it mean to be Complexity Conscious about innovation?

Accept that innovation is inherently uncertain. A healthy amount of variation and divergence is necessary if you want a vibrant ecology of self-renewal. Have the discipline to make bets in good times and bad.

WORKFLOW

How we divide and do the
work; the path and process
of value creation.

When Henry Ford introduced his moving assembly line in 1913, he changed the flow of work through the factory. The process was divided into eighty-four discrete steps, and none other than Frederick Taylor himself helped to optimize each of them. The build time of a Model T went from twelve hours to just two hours and thirty minutes. Nearly half a century later, Taiichi Ohno and the Toyoda family developed the Toyota Production System, a just-in-time approach to manufacturing that sought to eliminate *muri* (overburden), *mura* (inconsistency), and *muda* (waste). Once again, automotive workflow was forever changed for the better, and not just in the factory but throughout the entire organization. Improving workflow, the way value is created, is a continual source of advantage for the firms that do it. They gain speed, quality, efficiency, and in many cases simplicity. Yet it's routinely overlooked in favor of cosmetic changes to structure that rarely change how work gets done.

But wait a second, aren't structure and workflow the same thing? If we have clear roles, doesn't that tell us how we create value? Perhaps. Let's explore that through the lens of a coffee shop. For the sake of argument, let's consider just two roles: cashier and barista. The cashier handles the money. The barista handles the coffee. That's clear enough. But who will take the orders? The cashier could take them, charging customers while the barista works in the background. Or the barista could take them *and* make them while customers wait in line to settle up with the cashier. Both have their pros and cons. But regardless of which we choose, by establishing that accountability, we are *shaping* the workflow, how an order becomes an espresso. Behind that decision are a hundred more. Should the order be written down, called out, or entered into a system? How should the equipment be arranged in the space to support the process? If the line gets too long, what then? These questions must be answered, whether they're answered in agreements about roles or process or established as emergent norms through trial and error. Visit three different coffee shops. You'll find that they all have similar roles,

and they all make coffee, but each in their own way. Workflow, it would seem, is the fuzzy logic between our value-creation structure and our value-creation process. It's how the work flows through the organization. And it's how the work flows through the teams themselves.

Making a perfect cup of coffee is complicated, but it's manageable compared to making an Airbus A380, an aircraft with more than 2.5 million distinct parts produced by 1,500 different companies around the world. Just imagine *that* workflow and all the permutations of it that exist. How many projects are nested inside a product-development effort that massive? The answer, of course, is *too many* to keep track of. And this is true far beyond the world of aviation. In my experience, very few organizations have a handle on all their projects, how they move through the organization, or how they're managed day to day. This is, in many ways, a by-product of functional division. When you nurture large functions—engineering, marketing, human resources—you create an environment where important projects have to cut across the matrix and necessitate the participation of a wide variety of disconnected people. In terms of workflow, that's like salmon swimming upstream. So how do we make sure these people do what they're supposed to do? Enter *project management*. Since the team working on the project isn't really a team, we appoint a *project manager* to shepherd the herd. Of course, they lack the authority to really lead the effort, because all the participants' allegiances lie with their functions. Still, it gives us someone to hold accountable. And while we're at it, wouldn't it be nice if all these projects used the same "efficient" process? All hail the *project management office* (PMO), yet another group that will define and maintain the standards that our project managers enforce. This is what passes for workflow innovation in corporate America today—an additional layer of bureaucracy.

Of course, there are alternatives. Spotify, the Swedish music-subscription service, is often lauded for its structural innovation—the

squads, tribes, chapters, and guilds that have been copied (often badly) by other firms around the world. It's true that a network of multidisciplinary teams working on self-contained projects is part of the company's special sauce. But what's really impressive about Spotify's workflow is how it ensures that teams are "loosely coupled but tightly aligned." This concept, previously introduced in the Netflix Culture Deck (an employee handbook on steroids), suggests that we maximize strategic alignment but minimize dependencies and red tape among teams. When you're building a single piece of software used by seventy million paying subscribers, that's easier said than done. Early on, Spotify realized that if it treated its product as one monolithic application, the coordination required among teams would be too great. So the company cut it up and created the infrastructure to support a modular approach. Now teams own a specific piece of the experience (or the infrastructure) and have the autonomy to develop and deploy it independent of one another, in service of a continually renewed collective vision. Instead of one massive ocean liner winding its way through a series of functions, Spotify's workflow looks more like a regatta of speedboats heading in the same general direction. It's not perfect, but it's deliberate—prioritizing the things that matter to the company. May we all be so intentional.

Thought Starters

Flow Interrupted. If the way your teams are structured doesn't reflect the way your organization actually creates value, you're in for a world of hurt. Your sales team is waiting on your marketers, who are waiting on your product people, who are waiting on the engineers, who are waiting on infrastructure, which is waiting on funding. To witness this in action, choose a project that created real value for a customer. It could be a product, a service, or just a piece of one. Trace the line of who needed to be involved to create that value—by skills or by person, whichever is easier.

Now compare that network with the org chart that represents where people sit day to day. How many teams does it touch? What is the cost of those handoffs, in ownership, speed, and coherence? In an ideal state, your day-to-day value-creation structure—where people spend their time physically and virtually—and your workflow are one and the same. Your formal structure, to the extent you need one, is tucked in a drawer somewhere for the regulators.

Nothing but Projects. Could it be that an organization is nothing more than a series of projects? I know, the traditional definition of a project says that it must have a beginning, an end, and a defined scope and resources. But technically speaking, isn't that every activity we undertake? Everything from product development to cleaning the bathroom matches that description, depending on your time-frame orientation. Some projects last a century and some last a day. Just because we do something repeatedly doesn't mean it's independent of the bigger picture. Indeed, treating everything as a project allows us to be more intentional—starting, stopping, and evolving activity with the same level of scrutiny. A project has a purpose. A project has a cadence. A project uses feedback to learn. A project lens allows us to share a common language about the work of the organization. It forces us to take quotidian tasks and make them more strategic. Cleaning the bathroom as a *task* is disempowering. But the "clean bathrooms for our guests" project creates possibilities. Can we automate this? Can we design our bathrooms differently so they're easier to clean? Everything is in play. Of course, this doesn't mean tasks cease to exist. Just that they are often (and possibly always) in service of a project, consciously or unconsciously.

Workflow in Action

Working in Sprints. One way to establish a better rhythm at the team level is to work in *sprints*. Rather than basing your rhythm on the project's time line or complexity, consider running every project as a series of one- or two-week sprints. Think of a sprint as a unit of time during which you must produce *and share* a unit of work—a self-imposed deadline. While this might seem arbitrary, sprints enforce a lot of good habits that seldom emerge without a rapid cadence. For starters, they force teams to *get going*. If you know you have to ship something, anything, at the end of the first week, you don't waste time debating which #2 pencils to order. Sprints also force teams to break down the work into smaller bites. If you're building enterprise resource planning (ERP) software, or an app that scans brain waves, you won't be done in a week. So teams are left to figure out what *can* be done. An MVP. A paper prototype. A list of approved vendors. Because come Friday, *something* has to be shipped. And of course, sprints force decisions. Rather than debating or agonizing over what could be, the constraints mean it's safer to make a choice than to wait. If we show our users or customers that week's work and find out we missed the mark, so what? We spent one week to invalidate a direction that might have soaked up months of time in a previous regime. In my first meeting with a new project team, invariably someone will say, "When should we meet again . . . maybe three weeks from today?" And I'll reply, "What do you think we can ship by Friday?"

Limit WIP. Want to maximize how many cars pass through a stretch of highway? Pack it bumper to bumper and everything grinds to a halt. Instead, you want a constant flow of cars with enough space around them to keep moving fast. Limiting our *work in progress* (WIP) increases flow and overall productivity. But how does this work in practice? First,

think about how many projects you can reasonably juggle at once. Now cut that number in half. If it's more than seven, try again. Create a Kanban board with three columns: "To Do," "Doing," "Done." I recommend using Trello, but you can do this on the wall with tape if it suits you. Put everything on your plate in the "To Do" column. Now move the most critical and urgent projects over to the "Doing" column until you've reached your newly minted project threshold. When you finish a project, move it to the "Done" column. Then, and only then, you can pull a new project out of the "To Do" column and get started on it. Try this at an individual level *and* a team level. You'll be amazed how freeing it is. And remember, if a new request comes in that you agree to tackle, something else has to go back on pause. You can change a lot about your way of working, but you can't change the number of hours in the day.

Workflow in Change

As my colleagues and I begin to share and experience new ways of working with our clients, we're immediately confronted by their workflow norms. There's a lack of clarity about bandwidth because everyone is maxed out. As a result, there's an instinct to operate at a much slower cadence than we prefer. They think in big chunks of polished work; we think in small chunks of raw output. They think in months; we think in weeks or days. They coordinate through email, calendars, and assistants. We coordinate through messaging apps, cloud-based documents, and an established operating rhythm. Sometimes we try it their way, but more often than not we offer them the chance to try something new. And wouldn't you know it, most people quickly experience relief and a sense of momentum.

Questions on Workflow

The following questions can be applied to the organization as a whole or the teams within it. Use them to provoke a conversation about what is present and what is possible.

> > How do we divide the work of the organization?
> > What is the relationship between our workflow and our structure?
> > How do we handle projects that are too big for one team?
> > What is our approach to project management?
> > Who is accountable for project outcomes?
> > How do we maintain visibility across all our projects?
> > How are projects initiated, canceled, or completed?
> > What is the role of rhythm and tempo in our workflow?
> > How do we optimize our workflow to minimize waste and maximize value creation?

What does it mean to be People Positive about workflow?
Recognize that healthy workflow comes from organizing around the work, not working around the organization. When our teams and projects live in the same place, relationships fuel the work. And instead of pushing for uniformity, let local methods and tools flourish.

What does it mean to be Complexity Conscious about workflow?
Accept that workflow is something to be coordinated and refined, not something that can be solved. Ensure that every team has the capacity to do the work and improve how they do the work at the same time. In order to maximize the adaptive potential of the organization, create the infrastructure to support loosely coupled, tightly aligned teams.

MEETINGS

How we convene and coordinate; the many ways members and teams come together.

To meet or not to meet? That is the question on everyone's mind as we reconcile our visceral distaste for dysfunctional meetings with the fact that maybe, just maybe, we need them. Coming together around the fire as a tribe is an ancestral tradition that spans cultures and centuries. And there's no reason to believe that we've transcended that need. But we've allowed meetings to move well beyond their rightful place.

The average employee attends sixty-two meetings a month and considers more than half of those meetings a waste of time. The salary cost of *unnecessary* meetings is now $37 billion for U.S. businesses alone. Organizations aren't just overloaded; they're *addicted*. No daylight exists on their calendars, just back-to-back meetings as far as the eye can see. Their every need—for information, for decisions, for feedback—seems to call for a meeting. They hate them, but they can't quit, because there's no other way to get things done. Salesforce executive Vala Afshar masterfully highlighted the irony of our meeting culture when he tweeted, "You likely have to get management approval for a $500 expense . . . but you can call a 1 hour meeting with 20 people and no one notices."

Meanwhile, new tools have emerged that allow the more technologically savvy among us to communicate and coordinate across space and time. Teams that have adopted messaging apps such as Slack have cut meetings by 24 percent. And some have gone further than that. Fed up with meetings as usual, some renegade firms have banned them entirely, casting them out as wasteful and unnecessary in the future of work.

Are those our only choices? A wave of soul-crushing meetings or none at all? Pixar's Braintrust would suggest otherwise. The Braintrust is a part of Pixar's special sauce, the magic pixie dust that has enabled the company to produce nineteen number one box-office openings, fifteen Academy Awards, and an average Rotten Tomatoes score of 88.5 percent. The concept of the meeting is simple. The film, in whatever condition it's in, is screened in front of a handful of Pixar's most seasoned storytellers—directors, writers, heads of story—people who have been

there before. What follows is a two-hour free-form discussion, a feedback session known for what author Kim Scott calls *radical candor.* That's what happens when we challenge one another directly but make it clear that we care personally. The Braintrust leaves egos at the door, pulls no punches, and puts all their energy into making the movie better. As Ed Catmull, the cofounder and president of Pixar, says, "The film—not the filmmaker—is under the microscope."

There's one more thing that makes a Braintrust meeting different from a traditional feedback session. *The Braintrust has no authority.* The goal of the meeting isn't necessarily to *solve* problems but to shine a light on them, to trace them to their source, and give the creative team the perspective they need to start fixing them. Ultimately it's up to the director to decide how to proceed. "We don't want the Braintrust to solve a director's problem because we believe that, in all likelihood, our solution won't be as good as the one the director and his or her creative team comes up with." Teams leave the Braintrust with a renewed clarity about their film, one way or another, determined to unearth the greatness within.

Like it or not, great meetings offer higher bandwidth—more information per second—than any other form of communication. When we're together, we not only hear one another but also see and sense body language, emotion, and energy. Our mirror neurons start firing. We can shake hands. We can breathe the same air. We can stand shoulder to shoulder. Even organizations such as Automattic (the creators of Word-Press), GitLab, and Basecamp, which are famous for working remotely, host regular gatherings of their global membership. A million years of evolution didn't evaporate because someone invented the videoconference. If we want trust, if we want shared consciousness, if we want vibrant networks . . . eventually we have to meet.

Thought Starters

Death to Status Updates. One of the most popular and dysfunctional meetings I encounter in my work involves a team presenting work in progress to their boss for their feedback and/or blessing. Many leaders believe that these status updates are the best way to scale themselves across multiple projects. But holding court and dropping "pearls of wisdom" on teams actually causes more trouble than it's worth. For starters, leaders often lack the context to understand complex initiatives and the situation at the edge. This can lead to naive questions at best and irresponsible recommendations at worst. Updates are also disruptive, and not in a good way. Last-minute changes undermine teams and waste much of the time that was invested prior to the review. And, as you might imagine, these meetings become theatrical productions. Teams see updates as their chance to shine, so they overprepare, spending an average of four hours per week preparing for these moments. The alternative is far simpler. Leaders can either join a team and become part of the workflow, or they can be part of an advice process at the outset of the project or upon request. As one of my colleagues likes to say, "status should live in software." Words to live by.

One-on-One Fun. One meeting that employees love is the one-on-one they have with their manager every week or two. Most managers speak with pride about how often they meet with their direct reports and how "the time is theirs." We're servant leaders, right? Sure. But here's something I've noticed that you might not expect. One-on-ones are often used as a salve for hidden dysfunction. When members lack the authority to make decisions, these meetings become the only mechanism for moving things forward. When members lack the ability to resolve conflict, these private audiences with the leader become a forum for politicking. Great

one-on-ones can provide feedback and mentorship, deepen relationships, or give us a chance to collaborate on the work. But if you notice they're becoming a venue for other unmet needs, pull the rip cord and bring those conversations into the light.

Governance. Over the years, corporate governance has become a circus of compliance and risk avoidance. In the meantime we have forgotten that true stewardship comes from participation and ownership, not fear. Put simply, if we want organizations to learn and adapt—if we want them to be good citizens—then we need a distributed mechanism for steering and changing the organization. One way to do this is to encourage every team to hold a monthly governance meeting. The goal of this meeting is for everyone to have the chance to voice their concerns and propose local changes to structure, strategy, resources . . . anything that will help the organization pursue its purpose. Thousands of organizations around the world that practice Sociocracy, Holacracy, or other forms of participatory governance are doing this today. Imagine every team in your organization continuously and deliberately tweaking, not just your products and services but the organization itself.

Meetings in Action

Facilitators and Scribes. One of the best ways to increase meeting effectiveness is to ensure that someone is responsible for the structure, flow, and output of every meeting. Two roles that we have found to be particularly effective are *facilitator* and *scribe*. The facilitator role keeps the meeting on track, enforcing whatever format and ground rules the team has agreed upon. That could include cutting off conversational tangents, noticing when some people need to step up or step back, and even pointing out when the leader isn't playing by the rules. The scribe role captures any actions or outputs created throughout the meeting and

leverages any tools or interfaces required by the work. This could include anything from a digital Kanban board to a live document to a task manager. To get started, try electing a facilitator and scribe in each of your recurring meetings. Give them the authority to design and hold the space for meaningful interactions. After ninety days or so, elect new players to fill those roles, until everyone has held that responsibility.

Meeting Moratorium. Sometimes the only way to see things clearly is to stop the madness. Instead of trying to fix all the meetings in your current operating rhythm while they're in flight, see if your team is open to *canceling all meetings* for two weeks. At first blush, this sounds impossible, irresponsible even. But it can be done. One leadership team coached by The Ready had amassed an average of forty-five hours of meetings per week. Their calendars looked like a game of Tetris they were quickly losing. So we tried pressing pause on all recurring meetings. The questions we wanted them to answer? *What do we miss? What do we need that we're not getting from informal interactions?* Based on what we heard, we rebuilt the meeting rhythm one meeting at a time, ensuring that each one had a clear purpose and matching structure. We tweaked these formats iteratively, based on feedback, until we had something that worked. Gone were the needless reviews, cross-functional one-on-ones to negotiate and back-channel, and long-recurring meetings that no one could remember the impetus for. What used to take forty-five hours now only took eighteen. If you're overwhelmed, try your own meeting moratorium and discover the difference between what you *have* and what you really *need*.

Retrospectives. If you were to chart the most valuable but least practiced meetings, the hands-down winner would be something called a *retrospective*. A *retro* is simply a chance for any team to stop, notice, and learn. After a big push of work, or ideally on a regular interval, the team

will gather for an hour or two and share their perspectives on what happened, what stood out to them, and what they ultimately learned. The goal is simple: to do better next time. Many forms of retrospective exist, ranging from the simple (mapping highs and lows over the time line of the project) to the more complex (four Ls: liked, learned, lacked, longed for). The best kind of retrospective? The one that happens. Most teams have an abysmal track record of protecting the time for these sessions. Moving on to the next thing is too tempting. It's up to you to make sure that your team looks back and takes the time to learn regularly. It's also important to make sure that retros are a place where participants can *say what needs to be said*. If everyone holds back, the value of the meeting drops enormously.

Meeting Moves. While most organizations would benefit from fewer meetings, many would also do well to employ a wider variety of meeting structures. The vast majority of corporate meetings are a meandering mess, jumping from monologue to decision to idea to jest, wherever our stream of collective consciousness takes us. Instead, take the time to learn and employ proven *meeting moves*, which can be combined into meeting structures for unblocking the work, making decisions, providing advice, generating ideas, and so much more. Here are a few meeting moves that have added tremendous value to our meetings at The Ready.

Start meetings by checking in. Beginning with a question allows us to connect on a human level. We want to hear from everyone. Typical questions include: What's on your mind? What are you looking forward to? What's the biggest risk you've ever taken?

Speak and participate in rounds. When we want to prioritize speed *and* inclusion, we'll go around the table and give everyone *one chance* to pro-

vide updates, ask questions, offer feedback, or give consent, depending on the type of meeting we're holding. Everyone else is invited to listen respectfully and wait for their turn to speak.

Build an agenda on the fly. Rather than predict what will matter most tomorrow or next week, we'll choose and prioritize our topics once we get in the room. If we don't get to it, we don't save it. Someone will bring it up next time if it's still important.

If you're hungry for more, liberatingstructures.com is a fantastic resource for anyone ready to move beyond conventional meetings to something more inclusive and generative. The website, app, and the book that inspired them provide a menu of thirty-three methods for activities such as brainstorming, problem solving, and sensemaking, complete with instructions for how to facilitate them.

Meetings in Change

Because they're both so poorly run and the primary forum through which teams interact, The Ready's early work with organizations often starts in meetings. Small changes here can lead to breakthroughs in transparency, trust, time savings, and speed that create momentum and space for other, bigger moves. Perhaps more than anything else, meetings bring together the other dimensions of the OS into one shared experience. Don't hesitate to start here and use meetings as a microcosm of the bigger picture.

Questions on Meetings

The following questions can be applied to the organization as a whole or the teams within it. Use them to provoke a conversation about what is present and what is possible.

> What meetings do we require to do our best work?
> Does each of our meetings have a clear purpose and structure?
> How are meetings facilitated and documented?
> How are meeting outcomes and output shared?
> Which meetings are recurring and why?
> How does our meeting rhythm enable (or disable) the work?
> Do our meetings require any special tools or materials?
> How do we improve or eliminate meetings that are no longer serving us?

What does it mean to be People Positive about meetings?

Recognize that human beings crave connection and relatedness. Sharing the same space every once in a while matters. But rather than treating all meetings as social free-for-alls, let the purpose of every meeting dictate its structure. Some should work with our human nature, and some should help us transcend it.

What does it mean to be Complexity Conscious about meetings?

Accept that coordination and shared consciousness in a complex system require high-bandwidth forums for information sharing, including meetings. Remember that uncertainty applies to meetings too. Overpreparation and overgripping can lead you to miss what's present and critical for the team.

INFORMATION

How we share and use data;
the flow of data, insight,
and knowledge across the
organization.

When General Stanley McChrystal became the commander of the Joint Special Operations Command, an elite group of special operations forces including the Navy's SEAL Team Six and Army Rangers, he had his work cut out for him. Their mission was nothing less than to fight and defeat al-Qaeda, in Iraq and beyond. The problem was their opponent was unlike any they'd faced before. Decentralized. Networked. Empowered. Committed. And fast, very fast.

The military was none of those things. Hierarchy, secrecy, and security were in its DNA. Information was protected at all costs. After a series of challenging missions were thwarted by lack of information and speed in the field, McChrystal and his team had an epiphany. Their policy of "who needs to know" no longer worked, because *they didn't know* who needed to know. On more than one occasion they were *this close* to capturing or killing someone, only to find out that the target was working undercover and they were on the same side.

With everything on the line, they completely changed their stance on information. In a recent TED Talk, McChrystal shared what they learned. "What we found is we had to change. We had to change our culture about information. We had to knock down walls. We had to share. We had to change from 'Who needs to know?' to . . . 'Who doesn't know?' and we need to tell them, and tell them as quickly as we can." The JSOC literally did knock down walls, creating a new open command center that featured a wall of screens front and center with live updates of ongoing operations for all to see. They started including other leaders and teams on emails if there was even a *chance* the subject matter might impact them. They embedded liaisons across their network of agencies and locations. They transformed a daily videoconference called O&I (operations and intelligence brief) into a massive information-sharing session, inviting an unprecedented number of attendees, who were encouraged to speak with candor in a forum that had historically seen

little of it. JSOC's new guiding principle was "Share information until you're afraid it's illegal." By the end of McChrystal's tenure, every agency involved was reporting massive benefits from the new culture of transparency, and wins were materializing on the battlefield as well, including the death of Abu Musab al-Zarqawi, the leader of al-Qaeda in Iraq. For McChrystal and his team, getting information to flow was the only way to win. We too are trying to outsmart a network (the market) with a hierarchy (our organizations) and losing. And just like JSOC, we need to become intelligence-led.

Until very recently in human history, all information was trapped—in someone's head or, at best, on a piece of paper. This has led to a scarcity mindset when it comes to information. That, combined with a couple centuries of free-market competition, has made us wary of sharing. Within most Legacy Organizations, information is power. We hoard it to elevate our status, create job security, and protect ourselves from its misuse. Information is guarded and shared on a case-by-case basis. This perpetuates the power structure and creates opacity that allows bias and misinformation to thrive. Unchecked, this almost always leads to scandal and missed opportunities.

All living systems, at some level, are information processors. Starve any living thing of information and it will quickly perish. Professor Melanie Mitchell defines a complex adaptive system as "a system in which large networks of components with no central control and simple rules of operation give rise to complex collective behavior, sophisticated information processing, and adaptation via learning or evolution." In this way, a complex adaptive system is a kind of hive mind, capable of solving problems the individual agents cannot. Organizations are no different. Every single activity we conduct at work—recruiting, sales, accounting—is dependent on information processing and knowledge transfer. If we can tap into our *collective* intelligence, we can accomplish amazing things.

And so it's somewhat surprising how little time we spend on our information architecture—our approach to discovering, storing, and sharing what we know.

Data is not information. Information is not knowledge. Knowledge is not mastery. Mastery is not wisdom. These are important distinctions that can help shape our approach to individual and organizational learning. With deference to these distinctions, when using *information* as a dimension of the canvas, my intent is to include anything that can be transmitted or communicated. Everything from a raw data file to an email to a methodology to a conversation over coffee has a place in this dimension. If we intend to build vibrant, knitted, and networked cultures of learning, it all matters.

Thought Starters

Transparency. It's difficult to overstate the degree to which Evolutionary Organizations value and practice transparency. In complexity, as McChrystal and the teams inside JSOC discovered, insight can come from anywhere, but only if information reaches the right person at the right time. Because we can't predict when that will be, we have to aspire to what economists refer to as information symmetry—a condition in which all relevant information is known to all participants. This is a continuation and expansion of earlier concepts such as *open-book management*, in which employees have access to financial information such as revenue, profit, cost, cash flow, and expenses. The primary resistance to transparency stems from two beliefs: the Theory X notion that people can't be trusted and the scarcity-driven assumption that information is power. Today, companies such as Patagonia, Everlane, and Buffer are challenging all that. Patagonia offers public transparency about its supply chain through a video series called Footprint Chronicles that allows a customer to trace any item of clothing all the way back to its source—

from textile mills to sewing factories. Everlane goes one better. In addition to a transparent ethical factory network, on its website it shares the true cost behind each of its products, from materials to labor to transportation. The Everlane T-shirt I'm wearing as I write this cost $1.81 in materials, $5.60 in labor, and $0.13 in transport. And Buffer, the creator of a social media management platform, takes radical transparency to a whole new level, sharing everything with everyone. Its website (buffer.com/transparency) shows employee equity, salaries, real-time revenue, pricing breakdown, funding, values, reading lists, a log of every email sent, diversity metrics, their open-source code, the product road map, and the editorial backlog. Beat that.

Push vs. Pull. One reason we hesitate to adopt a more transparent approach to information is that we fear being overwhelmed. We're already swimming in a sea of messages and news feeds. Researchers at UC San Diego found that the average person is confronted with thirty-four gigabytes of information *every day*. The idea that we should all share everything all the time seems crazy. But that's only because we misunderstand how to share information—the difference between push and pull.

Legacy information sharing is "push," meaning that the information is delivered to us without our consent. When information is pushed, we have to wade through it and separate the signal (what we need) from the noise (what we don't). But when information is abundant, a "pull"-based system where information is tagged, stored, and ready to search is far superior. Email is push. The internet is pull. A single-track conference is push. A multitrack conference is pull. In the early days of Percolate, a content-marketing platform that was acquired by Seismic, it built a tool called Barista that allowed anyone in the company to ask a question and route it to people who might know the answer. Completed questions were tagged, saved, and searchable by

everyone else. Instead of trying to drown a new employee in pushed information, Percolate let them find what they needed when they needed it. When information is optional, accessible, and searchable, everyone wins. Less push, more pull.

Working in Public. When a culture is risk averse and teams lack the authority to make decisions about their work, a funny thing happens to work in progress. It goes underground. Why? Because teams know that if they share incomplete or imperfect work, leadership will likely poke holes in it and question their competence. So we end up with cultures where everything has to be perfect *before* it's shared. This leads to information silos that hurt everyone. Two teams working on the same thing won't know about it until it's too late. Leaders who might have helped shape early work in progress can only approve or deny. Anyone wanting to know the status of a project has to hope they know the right people and catch them on a good day. And anything and everything that could be repurposed and reused is trapped in place. Or we can *work in public*, which is shorthand for developing our work in applications and environments that are visible to and searchable by everyone in the organization. On day one at my company, a new member can search, find, copy, and reuse the active and prior work of the entire organization. Does that mean nothing is ever under lock and key? No. If you're legally obligated to keep certain information protected, you can adopt a policy of *default to open*. That means everyone starts with the assumption that all work should be open and searchable by default, unless there's a good reason to secure it.

Organizational Git. Even if you're not a developer, you may have heard of Git, an open-source version-control system created by Linus Torvalds that makes it easy for many people to work on the same piece of software

at the same time. The magic of Git is in how it enables distributed collaboration. Every project has a *master* repository of its code that is protected from reckless or duplicative changes. Contributors to the project create *branches* off the master that allow them to augment the code without changing the original. If their contributions are accepted, their branch can be *merged* with the master and available to everyone. Developers who want to create entirely new projects using an open-source master repository as their starting point *fork* the project, creating a copy they don't intend to merge. In this way, a single project can be improved continually and used as a basis for other projects around the world. It's the ultimate protocol for networked creativity. Now, imagine taking a similar approach to all the information, knowledge, and products in your organization. A master repository for customer insights. A master repository for product development. A mastery repository for your approach to feedback. And countless branches evolving in teams around the world, all with the potential to merge with the master for everyone's benefit. Imagine being able to search and borrow new ways of working from hundreds or thousands of other organizations. Curious about Patagonia's vacation policy? Why not *fork* it and give it a try? This concept is at the heart of murmur.com, the decision-making platform that we incubated and launched during the pandemic. The Ready and Murmur make most of our agreements public, so other organizations can use them too.

Information in Action

Kill Email. Despite decades of innovation, email is more popular than ever. Some 269 billion emails are sent and received every day, and that number is going up by 4.4 percent every year. The average employee checks their email thirty-six times per hour and receives 304 emails per

week. With all this email changing hands, we should have all the information we need at our fingertips, right? But that's not the case, not by a long shot. And the reason isn't that we're bad at email; the reason is that email is a completely inferior way to share information within an organization. It has three fatal flaws. First, email defaults to privacy rather than transparency. When you send an email, you have to decide in that moment *who needs to know*. Forget to include people who need it, and they'll be clueless. Throw caution to the wind and blast everyone, and you're wasting precious time and attention. Second, email is an information sinkhole. Sure, you might be able to find the email you need when you need it, but what about someone who never got it? What if a new employee joined your team today? Do they get the benefit of the thirty thousand emails buried in your inbox? What if you quit? What happens to all that information then? And finally, email is context free. Every email, regardless of whom it came from or how important it is, lands in your inbox in the exact same way. You want to know what it's about? *You have to read it*. If you're Seth Godin and you're sending your thoughts to hundreds of thousands of people, send an email. If you're trying to ensure information symmetry with a thousand trusted colleagues, you need a different approach. That's where messaging apps like Slack come in. These tools organize conversations into channels by topic, such as #marketing or #onboarding, and allow anyone to join or leave channels as they see fit. If you're looking for the status of your recruiting efforts, just go to the #recruiting channel and see what's going on. Conversations, files, and integrations with other software are all there in one place. You can still have a private conversation, or even a private channel, but that's a choice, and you have to ask yourself, Am I sure no one else would benefit from this discussion? Nine times out of ten the answer is no. Work in public. But here's the thing: messaging apps don't work if you keep email flowing. The inertia of what we know is just too strong. So here's what you do: get everyone's consent to ban email internally.

You'll still get email from the outside, but at the rate teams are joining Slack (it's the fastest-growing workplace software in history), you might find them in a shared channel before you know it.

Multiplayer Software. Traditional files are information traps—passed like hot potatoes and then lost in obscurity. If you're still exchanging documents with names such as "presentation-v32.7-final-ad-final-final .ppt," you're missing out on the cheapest productivity boost in the world: multiplayer applications. Apps such as Google's G Suite, Office 365, Dropbox Paper, Box Notes, Quip, Trello, Evernote, Basecamp, Asana, and Notion allow multiple users to create and edit documents, files, and data simultaneously. Instead of information flying from inbox to inbox, everyone shares a single source of truth that is always one click away. Teams can coauthor presentations, documents, and even entire projects synchronously or asynchronously, in the same room or remotely, for less than you spend on printer paper. When media company PopSugar switched to G Suite, the average time to go from an interview to a published piece went from twenty-four hours to less than two. Market research company Forrester found that organizations that switch to G Suite experience a 213 percent ROI over three years. Multiplayer software is now a prerequisite for efficient information flow. If you're already using it, double down. If you haven't tried it yet, start today.

Ask Me Anything. One powerful way to break patterns of secrecy and rumor is to host a regular *Ask Me Anything* session with your function, division, or organization. AMAs came into the mainstream consciousness via Reddit, the controversial online community and message board. Everyone from Barack Obama to Bill Gates to Jerry Seinfeld has gone on the site to be interrogated by the masses. An AMA works just like it sounds. Gather everyone together, physically or virtually, and start taking questions. Google famously includes an AMA at the end of

TGIF, its weekly all-hands meeting. At a credit union we advised, a weekly open-to-all coffee club provides anyone in the organization with a forum for asking questions or sharing thoughts related to the group's transformation. The CEO makes it a priority to show up, but he doesn't drive the meeting. The goal of these sessions is to tackle tough questions head-on, to show your organization that you trust them with challenging and sensitive information. In return, you can ask for accountability. Many organizations that hold real AMAs have nondisclosure policies around them to ensure that nothing gets out. And when it does (and it will), organizations that value transparency don't stop sharing. They simply fire the leaker and move on. Traditional AMAs often look to leaders for answers, but I've found that opening up the floor so that anyone can answer supports the idea that the best information is distributed. It also creates the space for more interesting conversations if different perspectives emerge in the room. Are you ready to find out what people are really thinking?

Information in Change

Increased transparency is critical in the early stages of transforming your OS, because it's a prerequisite for making sound decisions. One of the most common mistakes I see is teams taking a swing at empowerment before ensuring transparency. What happens? People make decisions without the benefit of crucial information (about intent, strategy, customers, prior learning, etc.), those decisions are subpar, and leadership goes, "See! People can't be trusted to make decisions." Avoid this by focusing on sharing early and often. Make it safe. Make it habitual. When shared consciousness is high, everything else gets easier.

Questions on Information

The following questions can be applied to the organization as a whole or the teams within it. Use them to provoke a conversation about what is present and what is possible.

> What information is shared freely?
> What information is contained or controlled?
> How do we decide what is safe to share?
> How is information stored and shared?
> What tools, systems, or forums support storing and sharing?
> How do we find the information we seek?
> How do we update our information when things change?
> What communication style(s) do we encourage?

What does it mean to be People Positive about information? Recognize that trusting people with sensitive information is worth the risk, even if the occasional breach occurs. Sharing leads to reciprocity, responsibility, and learning. Secrecy leads to distrust and suspicion.

What does it mean to be Complexity Conscious about information? Accept that no one knows what information will prove to be critical, or in which hands information might change everything. More and better information, and more and better ways to make sense of it, is the source of competitive advantage in complexity.

MEMBERSHIP

How we define and cultivate relationships; the boundaries and conditions for entering, inhabiting, and leaving teams and organizations.

Once a year, about a hundred miles outside of Reno, the Burning Man festival descends upon the desert for a weeklong social experiment in community, art, transcendence, and bacchanalia. Seventy thousand people supported by two thousand volunteers appear out of nowhere and stand up a self-contained society founded on ten principles: radical inclusion, self-reliance, self-expression, community cooperation, civic responsibility, gifting, decommodification, participation, immediacy, and leaving no trace. Simply by being there, by stepping into the space, you're making a commitment to those values. And if you're more than just a tourist, if you truly embody them, you're a "Burner." You're a member of the tribe.

The experience is so different from mainstream society that Burners refer to everything beyond their ephemeral city as the "default world." "The Burn," as they say, runs on a gift economy. You'll rely on the kindness and generosity of strangers, and they'll rely on you. Food and entertainment are not for sale, but there are plenty of both. People come together to form "theme camps"—subcultures and communal living spaces that host anywhere from a few to a few hundred residents. Joining a camp is like joining a team, and you'll be taking on duties and contributing to the camp's vision, whatever that might be. The experiences and entertainment that participants bring to life are almost beyond comprehension. Massive art installations, mutant vehicles, and elaborate parties celebrate the ideals and values of the community. You may find yourself. You may find God. You may run into Elon Musk (no relation). And the total cost to participate in this cashless economy where you'll be totally reliant on the people around you? It could be $2,000 or more. Tickets for last year's event sold out in just thirty-five minutes. This isn't Lollapalooza. It's not a few hours in the sun with your friends. This is a way of life.

Burning Man is a classic example of membership done well. When it comes to membership, it's helpful to think of an organization as a set of

membranes—cells within cells within cells. Each membrane, or team boundary, is made up of requirements and agreements, both spoken and unspoken. Honor them and you're in. Dishonor them, and you're out. The most traditional boundary, and the one we talk about most, is *employment status*. But there are other boundaries both within and beyond that. They exist around teams, functions, divisions, locations, social groups, interest groups, shareholders, customers, and even fans. Each one creates a communal space and identity for its members.

Legacy Organizations think about membership as binary—as a legal status or something that's conferred upon you. But membership isn't binary. Not every employee feels the same level of loyalty, or inclusion, or participation. No, membership is really a social status. It's an identity. It's a living agreement.

Boundaries can be clearly defined or purposefully blurry. Agreements within teams can be explicit or informal. Enforcement can be lenient or strict. What matters is that we are intentional. Evolutionary Organizations play with these continuums in an increasingly nonbinary way. Burning Man blurs the lines between attendee and host, customer and volunteer, and in so doing creates a richer and more participatory experience. Airbnb does the same. A host in one city is often a guest in another. Wikipedia does the same, as do countless open-source projects and peer-to-peer platforms. The future of membership may end up looking like many ways to play, clearly defined and held simultaneously.

As you may recall, in a complex system, the *interactions* matter more than the parts. A bench full of star players does not automatically produce a winning team. Relationships define our collective potential. And our approach to membership creates the foundation for relationships to flourish or flounder. Everything we do, from the moment we meet prospective candidates to the moment they depart as alumni, shapes their membership experience, and with it the very fabric of our network.

A member who is brought on board properly feels a sense of belonging and has a keen sense of how to move in and out of different membership spaces within the organization. A member who is thrust into a team or a role without that context may feel unwelcome and unclear. Do they belong? Are they safe? How should they navigate this system? Until resolved, a portion of their attention will always be dedicated to these questions. Recruiting, hiring, joining, onboarding, teaming, transitioning, disbanding, and departing—these are the domains of membership. These are mission critical. They are not the property of some faceless HR department on another floor in another building. Honoring and supporting these thresholds is our collective responsibility.

Thought Starters

In or Out. One of the reasons organizations end up rigid and risk averse is fear. Employees worry that if they don't perform well in their current role, they'll lose their job. And they're right: In the vast majority of cases, if your boss fires you from your role, you're also fired from your company. Your membership in the team and your membership in the organization are one and the same. But this makes sense only when we view the organization as a machine and our people as cogs. In that view, we have a fixed number of roles that require a fixed set of skills. If you're no longer the associate marketing manager, what the hell are you doing here? But if we view the organization as a living system and our people as multidimensional participants, then we can uncouple membership in teams from membership in the company. A member can hold one role, hold many roles, go find an open role, or even create a role. You can use an advice process or even elections to ensure that the people filling roles are competent. If the business enters a downturn, the organization might have to let people go. If a member's reputation severely impedes their

ability to join teams or make and keep commitments, they may have to go. But for the majority, who are sought after and have lots to give, the organization can be a jungle gym.

Teams Rule. W. L. Gore founder Bill Gore coined the notion of a "lattice organization" in which "each person in the Lattice interacts directly with every other person with no intermediary." In his view, "every successful organization has a lattice organization that underlies the façade of authoritarian hierarchy." In practice, this means that teams at Gore have to figure it out when it comes to teaming, and they attribute a lot of their success to the creativity and fluidity this enables. If you accept that each individual in your organization should choose their projects and their colleagues, something interesting happens. Teams become sovereign spaces and microenterprises. They have to generate their own resources, either through a budgeting process or through "charging" for their services. They have to cultivate their own membership, recruiting and removing members as needed. They have to develop their own norms and patterns of behavior. They have to create feedback loops, individually and collectively. They have to perform and add value to the broader ecosystem. This is the inherent tension between freedom and responsibility. Individuals and teams are free from hierarchical command but not free from constraints and accountability. They are in the roundabout, and everyone is counting on them.

Careful with Culture Fit. Organizational psychologist Adam Grant has some counterintuitive advice for those of us who hire for culture fit. While hiring for culture fit (even over skills or star potential) is a predictor of success early in a firm's history, later on it actually causes firms to underperform. In his research, he came across IDEO's approach to this issue. Instead of *fit*, IDEO hires for cultural *contribution*. It asks, What's

missing from our culture? and then goes and looks for that. Early on, we may need a handful of colleagues who see the world the same way in order to bring something new and disruptive into existence. But soon after, we'll need to focus on increasing the cognitive and general diversity of the group in order to achieve our full potential. We want our organizations to get more *interesting* over time. That's the kind of cultural complexity that helps us see around corners. In fact, a recent McKinsey study showed that companies in the top quartile for gender and ethnic diversity were 15 percent and 35 percent more likely to outperform those in the bottom quartile respectively.

Ritual. One way that we recognize membership is through ritual. Rituals help us mark important thresholds and changes in our lives, in ways large and small. The martial artist bowing as she enters and exits the dojo, the surgeon scrubbing up before entering the operating theater, the coming-of-age ceremonies present in so many cultures—these are all rituals that honor and mark transition. Research shows that rituals can reduce anxiety or increase confidence and even help us assume contextual identities (think: soldier or firefighter). In the world of sports, LeBron James welcomes each of his teammates onto the basketball court with an elaborate personalized handshake just for them. The level of memorization and coordination involved is baffling, but the meaning behind it is not. As team captain, he is recognizing each of his teammates and marking the threshold of the game. We practice a wide variety of rituals at The Ready, but one of my favorites is how we recognize team members who are leaving the firm for a new chapter in their career. We gather our team to mark the occasion, and anyone who feels compelled can share their gratitude for their departing colleague. A week later, our alumnus receives a globe selected just for them—a token of our appreciation and a reminder that we hope they'll change the way the world works wherever they go.

No Handcuffs. When you view the world as a zero-sum game in which there can be only a finite number of winners, you tend to view competition as something you want to eliminate. One way to do that is by restricting your employees with a legal instrument called a noncompete. Noncompete clauses prohibit employees from creating or joining competitive firms within a certain period of time after their departure from their current employer. Another way employers attempt to eliminate competition is through contracts and incentives that effectively bribe leaders to stay—until their contract expires or their stock options vest or their bonus is paid out. These moves illustrate confusion about human nature. Holding some-one hostage does not create peak performance. It creates resentment and disengagement. What if you didn't have these tools at your disposal? What would you do then? My guess is you'd focus on creating a workplace that people don't want to leave. I want my colleagues to *choose* to work with me *every day*. Online retailer Zappos feels the same way. They famously offer new employees $1,000 to quit after their first week or so. Why? Because they want only team members who are passionate about being there. Ama-zon, which acquired Zappos in 2009, liked this program so much they made it an annual offer within their fulfillment centers, upping the ante to $5,000 for longtime employees. Buurtzorg founder Jos de Blok goes well beyond abandoning noncompetes. When competitors come to find out how his company is able to do what it does, he actively coaches them on how to adopt his methods. Today Buurtzorg collaborates with approximately 80 percent of all Dutch care providers. Why? Because its intent is to increase health and wellness *overall*, not just within its clientele.

Membership in Action

Team Charter. In our rush to deliver, we often forget to do the founda-tional work of standing up a new team. Management puts eight people in a room and calls one of them the leader, and we're off to the races. This

is a missed opportunity. By tackling a few important topics early on, we can avoid a lot of confusion and conflict later. That's why I like to make the agreements that define a team explicit *before* we get started, by creating a team charter. Think of this a little bit like an OS for the team. A team charter forces the team to answer critical questions about why it exists and how its members want to show up for one another. To get started, gather a new (or existing) team and answer the following questions. Keep in mind, these are just a starting point. Real charters can go significantly deeper if the team is willing to do the work. When you're done, use a consent process (see page 70) to ensure that what you produced is safe to try. Revisit it at any time, particularly when there's disagreement or disruptive change that may impact the team.

Why does this team exist?

How do we contribute to the organization's success?

What are we accountable for?

What is our essential intent for the next _____ days?

How will we know if we've succeeded?

What principles will guide us?

What will we prioritize for the next _____ days?

What are the roles required to do this work?

What roles will we each play?

Are there any roles not yet claimed?

What do we expect of one another?

Who are our users or customers?

What decision rights do we have?

What can we do without asking permission?

Within what guardrails do we have autonomy?

Are we responsible for anything that we don't control?

How will we make decisions?

What resources do we control?

What is our meeting rhythm?

How often will we conduct retrospectives?

What tools will we use to communicate and coordinate?

How will we share our work with one another and the organization?

What are the learning metrics that will help us steer?

How will we know if we're making progress?

User Manual to Me. While chartering creates clarity about a team's overall purpose and context, it doesn't do a ton to nurture deeper and more functional relationships within the group. That's where the User Manual to Me, first introduced by Ivar Kroghrud in an interview with the *New York Times*, comes in and takes things to the next level. The idea is genius: what if we each wrote a user manual about how to work with us and shared it with our teams? Suddenly they'd know why we always seem skeptical or prefer to give feedback in person or get so excited about a good pun. How much time and confusion could we save? Adam Bryant, the former editor of Corner Office, who first interviewed Kroghrud, shared a list of questions that form the basis of a user manual at the New Work Summit in 2016. Ask your team to answer these questions in an easily accessible document that can be shared across your organization. Then bring everyone together, over drinks if you're in the mood for real candor, and have them each share their unfiltered answers. If you're the leader, go first. You'll be able to hear a pin drop, because you're about to share the secret code that everyone in the room has been trying to crack for months or even years: what makes you tick.

Questions About You

What are some honest, unfiltered things about you?

What drives you nuts?

What are your quirks?

How can people earn an extra gold star with you?

What qualities do you particularly value in people who work with you?

What are some things that people might misunderstand about you that you should clarify?

Questions About How You Relate to Others

How do you coach people to do their best work and develop their talents?

What's the best way to communicate with you?

What's the best way to convince you to do something?

How do you like to give feedback?

How do you like to get feedback?

Gratitude. One of the simplest ways to improve relationships within your community or team is to show appreciation for one another. Research shows that making time and space for gratitude improves well-being, reduces impatience, and boosts brain function. Here's an easy way to get started: At the beginning or end of your next meeting, ask everyone to stop what they're doing and think for a moment about something or someone they're grateful for and wish to recognize within the team. Then go around one by one and share. No fanfare, just an honest acknowledgment that says, "Hey, I love the energy you bring" or "You were there for me when I needed support" or "You're the best designer in the building and we're lucky to have you." A little touchy feely? Sure. But you won't believe the effect it has on morale. We forget to do this. We avoid it because it's uncomfortable. Don't let us. Make it happen.

Membership in Change

One of the most important steps The Ready takes in coaching teams through transformation is to define the boundary around who is actively working in new ways. This is critical because we are changing the requirements and agreements of that space in real time. That's why, as you'll see in the pages ahead, we focus on inviting (rather than forcing) people into a new way of working. We're telling them, if you step into *this* space, you're committing to try new things. That's what it means to be *ready*. Our membership is a coalition of the willing, of the zealous, of those eager to reclaim their way of working.

Questions on Membership

The following questions can be applied to the organization as a whole or the teams within it. Use them to provoke a conversation about what is present and what is possible.

> - What kinds of membership exist in and around your organization?
> - How is membership gained? How is it relinquished? How is it revoked?
> - What do all members expect of one another?
> - How are prospective members discovered and recruited?
> - How are new members brought into the community?
> - What is the nature of the relationships within and across teams?
> - How do members move between teams and other boundaries?
> - How are departing members carried out of the community?

What does it mean to be People Positive about membership?
Recognize that everyone needs to feel a sense of belonging,
both within the organization and within their team(s). Don't build
a walled garden that no one can escape. Ensure your boundaries
are porous enough for the membership to continually renew itself.
Celebrate generative difference and make space for people to
bring their whole selves to work.

**What does it mean to be Complexity Conscious about
membership?** Accept that a vibrant membership of commitment
and participation is a prerequisite for self-organization. Don't limit
yourself to the structures and policies of the past. It's unlikely that
an adaptive and resilient system is going to look like a traditional
employer filled with twenty-year veterans. Think about the
movements that inspire you. Model your membership after them.

MASTERY

How we grow and mature;
the journey of self-discovery
and development; our approach
to nurturing talent, skills, and
competence.

At Bridgewater Associates, the world's largest hedge fund, they have an unusual habit of recording their conversations. Every day, almost every meeting that takes place within the company is recorded. This is not some corporate urban legend. I know, because I've experienced it first-hand. A few years ago, I was invited up to Westport to meet with one of the company's executives. When I entered his office, he asked me point blank, "Do you mind if we record this?" Having heard about the Bridgewater culture long before I arrived, I was prepared for this. "Let's do it!" I said. He pressed a button on his desk—the room was wired—and we jumped into the discussion. Every once in a while, he would mark a moment in the conversation that he wanted to return to later. I'll admit, it was a little distracting at first, but I quickly forgot we were on the record. And while I thought I might be uncomfortable, I mostly left the meeting curious.

The fact that Bridgewater operates under surveillance wasn't what intrigued me. What intrigued me was *why* they did it—the principle of the thing. And the principle that drives Bridgewater is "radical transparency." Founder Ray Dalio and his team believe that if you want a culture that makes great decisions, you need a culture that is growing and learning all the time. So they record their conversations. They challenge one another. They "go back to the tape" the way a sports team might, to study what happened, how they came off, or exactly what was said. The Bridgewater culture is not for everybody (some might even call it cultlike), but the company's leaders credit its status as the most successful firm in the industry to their commitment to radical feedback. As I left their headquarters, I was handed a small spiral-bound booklet that contained a sort of blueprint for life and work. At the heart of it was this idea of feedback and learning. It was the rough draft of what would soon be Dalio's massive best seller, *Principles*.

Meanwhile, the rest of us are caught up in a completely different pursuit. "In an ordinary organization, most people are doing a second

job no one is paying them for," say Robert Kegan and Lisa Lahey, the authors of *An Everyone Culture*, "covering up their weaknesses, managing other people's impressions of them, [and] hiding their limitations." One of the consequences of weaponized Taylorism is that work has become a place to *perform*, not a place to *learn*. Confidence and equanimity get promoted. Humility, vulnerability, and struggle get labeled *weak*. So we put on a show for one another (and ourselves). Our egos become inflated and fragile. And learning becomes a secret shame. What utter nonsense.

Think back on the time in your career when you grew the fastest. What were the conditions that drove that rapid development? Let me guess. You were faced with a challenge. Perhaps you received some harsh feedback. Or you accepted an impossible project. In any case, you were in over your head. And you were no longer able to pretend that you were perfect. That freed you to ask for help, to try and fail, and to push yourself beyond your normal limits. And lo and behold, eventually you figured it out and leveled up. Not only did you know more about the challenge, but you knew more about yourself. We grow through resistance and discomfort. We grow when we are stretched, when the challenge before us exceeds our skills and we have no choice but to close the gap. But we can't appreciate those opportunities if we're too busy avoiding them.

Peter Senge suggested the concept of *Personal Mastery* in his landmark book *The Fifth Discipline*. He defined it as "a discipline of continually clarifying and deepening our personal vision, of focusing our energies, of developing patience, and of seeing reality objectively." This is strikingly similar to the ethos at Bridgewater, which was profiled as a "deliberately developmental organization" in Kegan and Lahey's book. One of Dalio's core principles is to "embrace reality and deal with it." In a company-wide email he asked his team, "Do you worry more about how good you are or about how fast you are learning?" What he's getting

at is that we have to get under the surface—and behind the masks we wear—in order to confront the truth and start to grow.

Practicing radical transparency is inherently uncomfortable. We will be exposed as imperfect. We'll have to face feelings and emotions that we normally avoid. We may hear things that threaten our egos or our sense of safety. Critical feedback from a trusted friend. Or an admission from leadership that funds are running low and the company may not make payroll. We may have to confront our privilege (or lack of privilege) and work through those complex dynamics out in the open with our colleagues. This is not easy stuff. This work requires that we show up at our best.

Here we see the connection between mastery and maturity. Talent and skills don't matter if we don't have the maturity—the courage and humility—to welcome the conditions for continuous growth. And our ability to do this has a lot to do with what psychologists call the *four dimensions of core self-evaluation*: locus of control, neuroticism, self-efficacy, and self-esteem. Our self-appraisal across these factors plays a predictive role in areas such as job satisfaction and job performance. Locus of control, though, is of particular interest as we explore mastery. The concept was developed by psychologist Julian B. Rotter to describe the degree to which people think they can control the outcomes of events in their lives. People with an *internal* locus of control believe that they have a high degree of influence over what happens to them. People with an *external* locus of control believe the opposite, that fate and other people shape their lives. Think of it as a personal referendum on determinism. Carol Dweck later expanded this concept into the world of learning in her book *Mindset*, which introduced the now-famous fixed and growth mindsets. Individuals with a fixed mindset believe their abilities don't change and their potential is predetermined. But those with a growth mindset believe that their effort and attitude determine their abilities. They see failure as an opportunity to grow.

Of course, what matters here isn't which perspective is *right* but which one produces the most beneficial outcomes. And in that context internal locus of control and growth mindset win, hands down. When you believe that your choices and attitude matter, you show up in your own life. You strive to make things better. You reach for your dreams. You never give up. This leads to more practice, more failure, more learning. Everything about the future of work calls us to connect with our deeper sense of self—our empathy, our vulnerability, our bravery, our humility, and our humanity. Ironically, it's the organizations that create the space to be *imperfect* that end up with the most stunning teams.

Thought Starters

Maturity Model Madness. One pattern in the learning and development space is to create maturity models for roles, skills, and methods that allow us to rate and rank ourselves and our colleagues. This is evocative of the belt ranking system—ranging from white to black—that emerged in martial arts in the early twentieth century (an auspicious time frame for putting things in boxes). Prior to that, rank was effectively just a private delineation between masters and apprentices. Why do we seek maturity? Because maturity is a state of self-sufficiency, competence, and judgment. That's wonderful. But maturity *models* stem from our misguided desire to cram all that complexity into a complicated framework. An ever-evolving spectrum of knowledge and skill becomes a fixed number of levels, each with its own qualifications. These models promote conformity to dogma and reductive criteria that can quickly become the focus instead of actual competence. In order to attain *this* rank you must know *these* things. A black belt in karate who has never been in a real fight meets a cage fighter with no formal training. Who is more likely to win? Who is mature? Luckily, it's easy to abandon the exhausting administration of these models for deliberate

practice and knowledge transfer between masters and apprentices. It's harder work, but we can stop chasing colored sashes and start getting good at what we do.

Learn by Doing. The other manifestation of our complicated approach to mastery is *training*. This is most commonly delivered by a "sage on the stage," an expert who imparts wisdom to a classroom full of students, but is increasingly done via web-based training modules that are the equivalent of a fancy PowerPoint deck. This approach to learning and development is still surprisingly popular—organizations in the U.S. continue to fund training to the tune of $93.6 billion a year. The only problem is information delivery and regurgitation is not learning, or at least not the kind we need. Anyone who thinks that a multiple-choice exam about unconscious bias is going to eliminate racism is kidding themselves. Today we face an array of complex issues, from privacy and security to sustainability to the future of democracy itself. The most in-demand jobs today barely existed ten years ago. Remember how the only way to understand a complex system is to interact with it? Well, the knowledge, skills, and mastery that matter most are being minted on the front lines, in direct contact with reality. That means our approach to training and knowledge management needs to change pretty radically. Dave Snowden, the director of the Centre for Applied Complexity at Bangor University, has spent decades thinking about knowledge management and ten years ago shared seven principles that challenge everything about our current approach:

1. Knowledge can only be volunteered; it cannot be conscripted.
2. We only know what we know when we need to know it.
3. In the context of real need, few people will withhold their knowledge.
4. Everything is fragmented.

5. Tolerated failure imprints learning better than success.

6. The way we know things is not the way we report we know things.

7. We always know more than we can say, and we will always say more than we can write down.

Each of those is worthy of its own chapter, but the takeaway is this: We need to accept that we cannot distill or transfer knowledge completely. We need to create work environments with high social density where members with different levels of knowledge and competence can work and learn together. We need to enable individuals to pursue roles and projects where their skills are stretched and flow is achieved. And when we do stop, we need to create safe and open spaces for experiential and self-organized learning, where memorization is replaced with dialogue and epiphany. Consider the case of Etsy, the online marketplace that hosts an internal Etsy School where the curriculum is taught by and attended by employees. Classes range from silk-screening to Python programming and help team members connect across disciplines in ways that help them grow in- and outside of work.

Mastery in Action

Role Mix. One of the greatest inhibitors to personal growth in the legacy workplace is the fact that we hold singular fixed roles. Being rooted in one function with a standardized set of accountabilities inhibits our ability to develop different skills in different contexts. It's not uncommon to work the same job for multiple years before a new role presents itself. This limits the potential of both the individual *and* the organization by reducing our ability to optimize the distribution of skills and interests against opportunity. In the structure dimension we talked about moving from one to many roles by atomizing our job titles. Shifting to a "role mix" enables us to pick up and put down roles as we grow.

And this leads to dramatically more interesting career paths. Instead of one grand promotion that changes everything, roles become a currency for learning and impact. People go where they will learn the fastest and contribute the most. To make this real for your team, start by asking everyone to articulate the role(s) they're *already* playing—including the role name(s), purpose(s), and general accountabilities. Have everyone evaluate which roles they want to continue to play and which ones they're ready to hand off. Next, give your team the responsibility to create, modify, and remove roles going forward. From here on out, every team member will be responsible for their own role mix, and every team for the roles within it. Of course, you'll have to decide whether roles are filled by appointment, election, or two-way negotiation, but that can be done on a role-by-role basis. And if you're starting to wonder how you'll compensate members in a system such as this, don't worry: that domain is coming up next.

Ritual Feedback. Traditional performance management is anything but People Positive, and it's incoherent when it comes to developing mastery. Reviewing someone's performance *annually* from a *single* perspective and *ranking* them with a number or grade . . . I mean, where do I begin? The time frame is woefully inadequate as a feedback loop. Anything said in that meeting should have been shared far sooner. What's more, reviews are extremely time-consuming. According to one study, the average manager reported spending up to 210 hours preparing for and doing appraisals annually. And worst of all, these processes make employees spectators in their own development. It's disempowering, and we've known better since McGregor argued for employee involvement in the 1950s. And of course there's the issue of separating individual performance from team performance that is inherent in processes such as this. Should a member of a subpar team who has strong personal performance receive a good review? What does that even mean? Many

organizations, such as Adobe, Gap, and IBM, have gotten hip to all this in recent years and abandoned the annual review process in favor of more frequent and inclusive processes. Real-time feedback from our leaders and colleagues is great, but there's also value in stepping back for a broader perspective. Here's an approach to balancing both that puts the person seeking feedback in control:

1. Set a time frame for deeper feedback that makes sense in your context. I recommend 120 days.

2. On that interval, using software (e.g. Slackbot or Google Forms) or a team willing to facilitate, reach out to each member of your team, and ask them if they'd like feedback. If they say no, let it go. They'll get another chance in a few months.

3. If they say yes, ask them whether they'd like to use standard questions (start/stop/continue or similar) or write their own.

4. Ask participants which three to five colleagues they'd like to receive feedback from. If you're using software to automate this process, you may even be able to recommend the people they communicate with most frequently.

5. Send the questions to their suggested colleagues with a time limit for contribution. Make this a ritual that is prioritized and celebrated in the culture.

6. Compile and share the responses with each participant. Let them choose who to share it with, including their manager (if they have one).

7. Invite participants to convene their respondents for an in-person sensemaking session led by them. The question on the table is "Based on the feedback you received, what questions do you have for your colleagues?" This is chance to zero in on things that they want to work on and make real breakthroughs and commitments.

8. Finally, don't let this feedback rhythm become an excuse to avoid real-time conversations and retrospectives. The end of every interaction, sprint, and project is an opportunity for feedback. Teams at The Ready try to take a moment for feedback after any key event. Leaving a new business meeting, I might ask a colleague, "What did you notice? What happened? What surprised you? And what feedback do you have for me?"

Communities of Practice. If engineers don't sit with engineers . . . if senior marketers don't manage junior marketers . . . how will they learn their craft? This is common question that surfaces when we move away from functional silos to functional integration. The answer is pretty obvious: members who share a common practice or interest should get together regularly with the express purpose of sharing and learning from one another. If our structure is optimized for value creation, then we don't want UX designers spending every waking moment together, but we don't want them to be strangers either. Encourage the organic formation and stewardship of communities of practice or interest. Spotify, for example, calls groups dedicated to a specific discipline, such as software testing, *chapters*; groups that share a common interest, such as sustainability, *guilds*. Each one meets on its own rhythm and format, and each one is facilitated by one or more members who have expertise in that area and/or feel called to community. Moving in this direction is as easy as sending a calendar invite:

Sustainability Meetup / THURS @ 4pm / Come if you want to chart a path for how we can go carbon neutral in the next 90 days!

UX Community of Practice / WEDS @ 9am / Anyone with a role or interest in UX come and share questions, learnings, and advice!

Mastery in Change

A certain level of maturity and competence is required to participate effectively in an Evolutionary Organization. Members who lack self-awareness or self-confidence may struggle with self-management. If you're not able to be vulnerable, radical candor can be too threatening. It's a shock to the system even for those of us who think we've mastered our egos. If you're not clear on your own talents and purpose, self-direction can seem overwhelming. *Everyone's always told me what to do, and now I have to decide?* That's why it's important to view OS change as a continuous unfolding. One practice leads to another, and we build mastery through repetition. Start where you are, not where you want to be.

Questions on Mastery

The following questions can be applied to the organization as a whole or the teams within it. Use them to provoke a conversation about what is present and what is possible.

> - What is our approach to learning and development?
> - How do we define and assess competence?
> - How does competence influence the roles we inhabit?
> - What knowledge and skills are required to pursue our purpose?
> - How do we give and receive feedback?
> - How do we spread competence and mastery within our membership?
> - How does competence shape or influence our career paths?
> - What is expected of members in terms of learning and personal growth?

What does it mean to be People Positive about mastery?

Recognize that developing personal and professional mastery is a basic human need. If you can create an environment where people grow quickly, you'll never lack for stunning colleagues. People are the architects of their own growth. Let them solicit and guide their own feedback and progression.

What does it mean to be Complexity Conscious about mastery?

Accept that competence is complex and contextual. Don't try to reduce talents and skills to a matrix. Remember that reverence for expertise can lead to arrested development or missed opportunities. Learning agility is a far better bet. A great learner can figure it out on the fly. The world's best hammer can only hit nails.

COMPENSATION

How we pay and provide;
the wages, salaries, bonuses,
commissions, benefits,
perquisites, profits, and equity
exchanged for participation
in the organization.

Everyone needs to eat. Perhaps that's why the word "salary," from the Latin *salarium*, is so similar to *salarius*, the Latin word for salt. Work is the means by which we meet our basic needs: food, water, and shelter. The modern concept of compensation goes well beyond our basic needs, though. It has evolved from simple wages into a diverse collection of financial and nonfinancial instruments intended to attract, retain, and even motivate a wildly diverse workforce. Despite decades of evidence to the contrary, organizations still treat compensation as a motivator. Want to make someone happy? Pay them more. Want them to work harder and hit their targets? Dangle a carrot in front of them.

A well-positioned full-time employee in corporate America is likely to have a base salary that is paid to them once or twice a month, a benefits package that includes health insurance and a 401(k) retirement savings plan, paid sick days, paid time off, and a performance bonus. An in-demand employee at one of the top firms in the world is likely to add to that list a combination of stock options, extended parental leave, child care, transportation, free meals and snacks, expense-paid holidays, corporate retreats, professional development, conferences, sabbaticals, meditation rooms, on-site exercise, unlimited time off, the option to work remotely, free technology and supplies, corporate housing, and of course the office ping-pong table. In the war for talent, the modern workplace looks more and more like an all-expense-paid trip to Club Bureaucracy.

Lacking purpose and meaning in their work, many professionals view their career as nothing more than a series of stepping-stones from one job title and pay package to another. But this frenzy on the part of both employers and employees betrays a misunderstanding about the connection between compensation and outcomes such as satisfaction, happiness, and loyalty. Several studies have attempted to examine the link between greater pay, incentive pay, and corporate profitability, along with other measures of performance. What they found amounts to *no or an even slightly negative correlation* between pay and results.

Management guru W. Edwards Deming had it right when he said, "Pay is not a motivator." But then what is it? In 1959 psychologist Frederick Herzberg proposed an answer to that very question in what would come to be known as his two-factor theory. In trying to understand the factors that lead to job satisfaction, he discovered that the opposite of satisfaction was not dissatisfaction. Rather, satisfaction and dissatisfaction appeared to be driven by two completely different sets of variables. *Motivators* included job characteristics such as recognition for achievement, meaningful and interesting work, involvement in decision making, and advancement or personal growth. Addressing these factors increased satisfaction by improving the nature of the work itself. *Hygiene factors*, on the other hand, included company policies, job status and security, supervisory practices, and—you guessed it—salary and benefits. Addressing these factors *reduced* dissatisfaction by improving the job environment. What does this mean for the world of compensation? It means that increasing salaries that are too low can reduce job *dissatisfaction*, but increasing salaries that are already generous won't increase job *satisfaction* in any meaningful or lasting way.

It won't make us any happier either. According to a study published in the *Proceedings of the National Academy of Sciences*, household income above $75,000 a year doesn't make us measurably happier. Below that level, increases or decreases in income were correlated with happiness. Daniel Kahneman, one of the pioneers in the science of cognitive biases, was one of the authors of the study. In an interview with the *New York Times*, he commented, "It's not so much that money buys you happiness but that lack of money buys you misery." Sound familiar? Herzberg was right. This is not to say that every job should have a maximum salary of $75,000. Supply and demand still play a part in shaping the market for any given role. But it should give us pause as employers attempting to pay our way to performance and satisfaction, and as employees weighing the merits of different offers and career paths. As Jack Antonoff of Bleachers

urges in his 2017 anthem of the same name, "Don't Take the Money." Compensation is hygiene. Don't mistake it for higher purpose.

Thought Starters

Parity Problems. It's been a half a century since the Equal Pay Act, but women still face a sizable wage gap, earning somewhere between 78 and 80 cents for every dollar a man earns. Many minorities face similar gaps, which are in some cases actually increasing. These pay gaps are a complex phenomenon, driven in part by discrimination and bias but also by broader educational and sociological patterns. A new study on the gender pay gap in Denmark, one of the world's most egalitarian societies, reveals that women working full time are still paid 15–20 percent less than men. Surprising, until you realize that pay gap is largely a penalty for childbearing. Women experience a sharp decline of up to 30 percent in their earnings after the birth of their first child. This is less about different pay for the same work than about choosing different work altogether, seeking greater flexibility and/or reduced hours to accommodate a different set of demands on their time and energy. While their earnings continue to grow at a pace similar to men's, women never fully recover from this dip. This phenomenon also manifests in what is now a well-publicized leadership gap. Women make up 44 percent of the overall S&P 500 labor force, but only 25 percent of executive and senior-level positions and 6 percent of CEOs. Any organization seeking to provide equal pay or equal opportunity today is facing a challenge more to do with privilege than with pay scale.

Formulaic Pay. One response to bias and favoritism is formulaic pay, an approach that organizations such as Buffer and Stack Overflow have publicly adopted. By visiting their websites and entering a few variables such as the role you're considering, your level of experience, and your location,

you can calculate the salary you'd earn if you worked there. By design this is a take-it-or-leave-it offer. Once you've gone through their recruiting process and your variables are firmly in place, there's no room for negotiation. You'll be paid the same as everyone else at your level. This is how the formulaic approach reduces bias; historically marginalized people negotiate less aggressively than their white male counterparts. This can be challenging for candidates used to pushing for more, but it's a philosophical commitment that doesn't work with exceptions. Another offshoot of the formulaic approach is contribution-based pay. In this model, peers rank or evaluate one another's contribution (or ability to contribute). The resulting distribution of people is mapped algorithmically or by committee to a set of salary bands or levels. Whether this results in fairness or a popularity contest depends on the maturity of those involved—but then again it can be hard to parse the difference. Both these approaches run the risk of oversimplifying an inherently complex topic. How do we know seven salary bands is the right number? And is it possible to balance quality of life by calibrating salaries to cities? These and other questions taunt us.

Market Pay. In the spirit of self-regulating systems, many organizations base their compensation on the market. Approaches vary, but essentially this means using the industry average salary or wage for a particular role as the baseline for your own compensation decisions. Netflix famously takes this approach to the next level by paying *top of market* for the person in question. This is an individualized approach also known as "paying the person." It uses three questions to determine the top-of-market value for a member of their team: (1) What could this person get elsewhere? (2) What would we pay for their replacement? (3) What would we pay to keep this person, if they had a bigger offer elsewhere? The goal is to consistently keep each employee at the top of their own market value. Sometimes that means large or frequent raises. Sometimes that means reduced pay. Unlike a typical market-based model, Netflix is

actually skeptical of using titles or roles to determine the market, as "all people with the title 'Director of Engineering' are not equally effective." Paying top of market is great, but in many cases it's still dependent on a manager interpreting someone's value from their singular perspective. Unconscious bias and favoritism are bound to creep in.

Self-set Pay. If formulaic pay is too reductive and market-based pay is too preferential, what's left? Letting employees set their own salaries. This radical notion captured the imagination of a handful of early adopters that have been doing this for decades. At Morning Star, the tomato-processing company I mentioned earlier, you'll remember that employees take part in an annual ritual where they write their own job descriptions and set their own pay. Using internal and external compensation data, the firm's financials, and their role mix, they craft a recommendation for their salary or wage for the year ahead. Next an elected compensation committee made up of their colleagues reviews these proposals and provides feedback in the form of advice. Some employees are advised to soften their ask—perhaps their track record doesn't support such a big raise, or the market is softening and the firm needs to be careful with expenses. Others are encouraged to reach higher; they may have undervalued their contribution. The advice is just that—guidance that employees can heed or not. Most do. And a conflict-resolution process does exist for any extreme outliers, but this is rarely needed. The transparency around compensation is sufficient for most people to make sound decisions that can withstand the scrutiny of their peers. As eccentric as this sounds, it's not *that* uncommon. In addition to Morning Star, organizations such as Semco, Bruggink & Van der Velden (BvdV), Incentro, Hanno, elbdudler, Premium-Cola, and AES have each allowed employees to set their own salaries without breaking the bank. But beyond that, the benefits in financial acumen, stewardship, and collective responsibility that this approach produces are unparalleled.

Gig Economy. The platforms behind the gig economy like to talk about their movement as the savior of the American worker, empowering otherwise underemployed individuals to be their own bosses and live the entrepreneurial dream. After all, the drivers and laborers who make Uber, Lyft, Grubhub, DoorDash, Postmates, Fiverr, and TaskRabbit work can choose when and where they work with unprecedented control. Realistically, though, many of the workers in the gig economy need money. That's why they're side hustling. They're underemployed or unemployed, and the minimal extra income they earn from these services—85 percent make less than $500 a month—is helping them make ends meet. That doesn't sound like the ultimate in entrepreneurial freedom. But there's something more troubling about the fact that one in four Americans is now participating in the gig economy. By turning work into a series of app-mediated transactions, we're actually narrowing the scope of their participation to something closer to the opposite of entrepreneurialism. When you work at Lyft full time, you're (hopefully) looking for ways to grow and serve Lyft all the time. If you see something worth doing, you might just do it. But when you *drive* for Lyft as a gig, your relationship is read-only. You transact, but you do not serve the bigger picture. Why would you? And that's the problem. If we move toward an economy where everyone is paid "by the drink," we run the risk of eliminating good corporate citizenship. If we *thin slice* the work too much, we'll watch as "that's not my job" becomes a mantra and a way of life. And that disconnection will rise at precisely the moment when we need all hands and minds on deck to invent the future.

Compensation in Action

Transparent Compensation. Considering the significance we attach to compensation, it's somewhat surprising that the subject is taboo. At least in the Western world, we don't talk about money—it's actually considered

distasteful. Talking openly about how much we make, either in public or within our teams, is less popular than talking about politics. Of course, this deference to polite society has real societal consequences. The main by-product of this secrecy is "information asymmetry," which is what happens when one party in a negotiation has more information than the other. Whoever knows less is at a disadvantage. In this case, employers know a lot (such as what every single employee at your level is making), and employees know very little. This can result in a wide range of salaries for the same role, known as wage dispersion, both within firms and across industries. More than half of this dispersion can be attributed to workers not knowing the wages offered by all employers and cross-firm differences in wage policy and productivity. One easy solution to all this is to abandon the taboo around salaries and make them transparent, both within the firm and to the broader public. Buffer, in addition to sharing its salary formula, also shares the salaries and equity of everyone at the firm on its website at buffer.com/transparency. And before you remark on Buffer's relatively small size (~120 employees), keep in mind that Whole Foods has shared salary data internally since 1986. In fact, dozens of organizations including Namasté Solar, CareerFoundry, Crowdfunder, SumAll, and the federal government practice internal transparency when it comes to pay. They report that although it can soak up some additional time and energy, the results speak for themselves: more job applicants, reduced bias and unfairness, and increased trust. Some studies have even shown that environments with transparent pay are measurably more productive. It may still feel like a radical choice, but websites such as Glassdoor and Salary.com are giving employees a platform to share anonymously right now. Glassdoor, for example, has more than 40,589 user-contributed salaries listed for Microsoft in the United States alone. The cat's out of the bag. Transparency is coming. If you want to get ahead of it, here are a few steps to get started:

1. Ensure that current salaries reflect your own definition of fairness and that you have a rationale or a formula that explains the status quo. (Because you're going to get a lot of questions.)

2. Start with general education and discussion about the company's finances and approach to compensation. Make sure everyone has a chance to understand the theory behind the current approach.

3. Assess whether your team has the comfort and confidence to voice any concerns that might be raised by freshly transparent salaries. The more open the culture is, the more likely you'll weather the initial bumps and bruises of opening up.

4. Ask for consent from everyone involved, and decide ahead of time if you're looking for total buy-in or majority rule. This is an extremely contentious topic, and you run the risk of losing people if you push ahead irreverently. It's okay to start with one division or unit that's willing to charge ahead and prove the model.

5. Publish salaries in an open and easily accessible forum, and convene a combination of all-hands and active-team meetings to create space for discussion and sensemaking. Encourage anyone and everyone to speak up if they see inequity in the compensation system. Be prepared to make adjustments, provide counsel, and say good-bye to a few people who may not be able to metabolize the new world order.

6. As the volatility of the initial announcement wears off, ensure that your approach to continual salary adjustment is compatible with your newly transparent culture. Assuming it is, sit back and enjoy the benefits of living open!

Eliminate Bonuses. Incentive compensation is counterproductive. At best, it rewards behavior that was already happening and, in the process, strips that activity of its intrinsic value. At worst, it actually promotes

negative behavior in the form of sandbagging, unhealthy competition, and manipulation. To make matters worse, individual performance is a myth. Not because there aren't heroes among us capable of amazing feats—there probably *are* software developers that code ten times better than the mean. But being a great player is not the same as winning the championship. Just ask the 2004 U.S. men's Olympic basketball team. A team consisting of MVPs such as Tim Duncan, Allen Iverson, LeBron James, Dwyane Wade, and Carmelo Anthony *lost* . . . to Argentina. Why? Because they lacked chemistry. True performance—the kind we ultimately care about—is a team sport. So instead of taunting people with bonuses, pay above market for the best talent you can recruit, and get out of their way. If you really want to reward your team for business performance, do it in the form of profit sharing proportional to their percentage of the salary pool. That metric is harder to game, and it brings everyone together.

Compensation in Change

Because compensation is often opaque and connected to roles and titles, the work of reinventing the OS and shining a light on what's really going on can provoke anxieties about this dimension. Those who benefit disproportionately from the status quo may be hesitant to be exposed, either out of guilt or for fear of corrective measures. Those at a disadvantage may not even know, but confronting the facts may make them feel undervalued or underappreciated. While not always possible, it can be helpful to provide some downside risk protection for everyone, such as "while we adopt pay transparency and new approaches to compensation, no one will cut your pay without your consent."

Questions on Compensation

The following questions can be applied to the organization as a whole or the teams within it. Use them to provoke a conversation about what is present and what is possible.

> - What is our approach to compensation?
> - What other benefits or services do we provide?
> - What incentives, if any, do we offer?
> - How do we define and ensure fair compensation?
> - What mechanisms have we put in place to reduce bias in compensation?
> - How are changes in compensation triggered and conducted?
> - Do we offer profit sharing or equity compensation?
> - How do our approaches to membership and compensation interact?

What does it mean to be People Positive about compensation?
Recognize that compensation is a hygiene factor that should be fair and generous enough to not matter. Keep the focus on autonomy, mastery, and purpose—conditions that actually support motivation. If you can, move to profit sharing (or a similar construct) to connect everyone to the overall success of the organization.

What does it mean to be Complexity Conscious about compensation? Accept that no formula, leveling system, skills matrix, or series of job titles is going to sufficiently capture the complexity of a real workforce. Only transparency, dialogue, and judgment can make sense of what is fair, and even then you'll be grappling with inherent issues of bias and privilege that will not be easily eliminated. Compensation can't be solved, it must be tuned.

USING THE OS CANVAS

Now that you're familiar with the dimensions of the OS Canvas, let's briefly talk about how to use it. Like many tools, the canvas can be leveraged in a variety of different ways. At The Ready we've seen teams use it *descriptively*, to articulate their way of working (or someone else's). We've seen teams use it *diagnostically*, to explore a positive or negative pattern they've noticed (e.g., why do new hires feel so confused about our onboarding process?). And some use it *aspirationally*, to imagine ways their firm might evolve.

For our part, we most often use it as a sensemaking tool—to map stories, tensions, and experiments that are happening in the wild— asking the teams themselves to interpret what's going on. Regardless of method, the canvas has a tendency to produce epiphanies in teams as they begin to think systemically about their way of working. A meeting isn't just a meeting. It's a forum for membership. A chance to share information. An opportunity for consent. Or it's a waste of time. The canvas can provoke that conversation, and that conversation can provoke a change.

READY OR NOT

All decisions involve emotion. When you face a choice, subcortical structures in your brain fire and a range of emotions, instincts, and body sensations unfold, influencing or even arriving at a somatic decision before you're fully aware of it. This is what is commonly known as a "gut feeling." This interplay between our various thinking systems is so fast and embedded that we don't even realize it's happening. We may cling to the idea that we make most of our decisions objectively and rationally. But it simply isn't so.

Which is why, at this point in the book, I believe you've already made your choice. Somewhere in your mind or your body, you know. Either you believe that we need to change how we work, and you're going to take that leap, or you don't and you won't. From here on out, I'm speaking to the believers. The catalysts. The visionaries. The risk takers. The readers who know *in their gut* that if we don't change how we come together as human beings to build and shape the future, then the future won't be a place we want to be.

So where do we go from here? The answer is easy; the doing is hard. If you hold power over other people—in business, in philanthropy, in education, in public service, in your community, or even at home—it is your responsibility to increase the humanity, vitality, and adaptivity of that system. If you're with me, here we go.

Part Three

THE CHANGE

> Beware that, when fighting monsters, you yourself
> do not become a monster . . . for when you gaze long
> into the abyss, the abyss gazes also into you.

—Friedrich Nietzsche

HAVING HAD AN EPIPHANY ABOUT THE FUTURE OF WORK, IT IS tempting to march back into the office and announce that *things are going to change around here*, even before you read on. Because that's what leaders do, right? They inspire us. They push us into the unknown.

Resist that urge. In fact, do the opposite. Stop.

There is something paradoxical about what you're about to do. You're trying to lead your organization to a place where you're *not the leader anymore*, at least not in the way you are today. How do you get people who have been managed all their lives to suddenly self-manage? How do you get a culture that is addicted to planning and control to realize there are better ways to manage risk? How do you get leaders whose identities and egos are wrapped up in status and position to realize that this power is not the source of their value? These questions and dozens more are going to confound you.

In the coming pages, I'll share a straightforward process for safely and continuously evolving your OS. But first, if you really want to change— if you really feel compelled to reinvent your way of working—then you're going to need a lot more than a process or a checklist. You're going to need to forget everything you know about how cultural transformation happens in organizations.

THE CULTURE CONUNDRUM

Faced with so many radical and exciting possibilities, you're probably contemplating how to change your *culture*. Yes, culture, that most powerful and misunderstood word that beckons like a siren. It's at the beginning and end of every deep conversation about organizational success, isn't it? Everyone says, "Culture eats strategy for breakfast." More potent than strategy, and yet still so elusive. It's the one word in business that means everything and nothing. Author Seth Godin defines it as *the story we tell ourselves*. "People like us do things like this," he says. It's that simple, and that hard.

The role culture appears to play in success and failure has led to a widespread belief that it's something we can and should direct— something we can change. But that's a fundamental misunderstanding of what culture is. Culture can't be controlled or designed. It emerges. It isn't happening *to* people; it's happening *among* people. Author Niels Pflaeging explains the inscrutable phenomenon with a powerful metaphor: "Culture is like a shadow: You cannot change it, but it changes all the time. Culture is read-only." Nevertheless, every year leaders valiantly attempt a miracle: to fix or change their cultures with PowerPoint and promises. Spoiler alert: these programs don't work.

Several years ago, I consulted as part of the team that refreshed the values (known internally as "beliefs") at a global industrial company. The values reflected what we, a team of fewer than ten, believed was

right for the other 300,000-plus people who worked there. The output was just five sentences, but the rollout of those sentences took *years*. Countless hours and millions of dollars were spent trying to encourage employees to *live the values*. In the end, did everyone "stay lean to go fast," as one of the values suggested? Not as much as we would have liked. And I don't blame them. A platitude is not an epiphany. We can't change hearts and minds with a haiku.

When culture proves too amorphous, we turn to simpler adversaries. Leaders say, "Our people are the problem!" And the people say, "Our leaders are the problem!" And so we attempt a second miracle: changing one another. But again, this is a misunderstanding of human nature. People are complex. We grow and change in our own way, in our own time. As Ben Franklin once said, "Consider how hard it is to change yourself and you'll understand what little chance you have in trying to change others."

So, if we can't change culture and we can't change people, what *can* we do? We can change the system. We're headed into uncharted territory now. Here there be dragons.

Control Inc.

We were three weeks into a project and our sponsor on the client side was looking for *the plan*.

"I just don't understand what's going to happen," he told me. "We're meeting with teams and talking about their way of working, but when are we going to start doing the real work?"

"Well," I said, "this is the real work. We're asking teams what's slowing them down or standing in their way. We're asking them what they'd change if they could. And we're inviting them to design and run experiments around those tensions and scale the ones that work."

He looked at me quizzically and took a long pause.

"Right . . . but what's *the plan*?"

His company, Control Inc., had reached out to us because they were seeking a partner to help with culture change. The leadership team was under pressure to turn the company around because growth had stalled and employee engagement was dwindling. They had rightly identified that culture was central to their predicament. However, the other partners they were considering—traditional consultants and Agile experts—made us think twice. Those were the kind of partners you chose if you wanted a concrete recommendation. A PowerPoint deck. A proven methodology. A structural reorganization. These sorts of solutions can feel comforting when you're unsure, but over time they lead to dependence and fragility. When the one-size-fits-all answer doesn't work (or stops working), what then? We made the case that what Control Inc. needed was bigger than Agile and more nuanced than a reorg. It needed to evolve its operating system.

To our surprise, our pushback won us the project. But since the kickoff, there had been several "lost in translation" moments that made us question our alignment. This latest request for the plan was just the most recent example. It was all the more frustrating because we had really struggled to get dedicated time with leaders and teams to begin the work. We wanted access and volunteers; they wanted a plan.

"Say more," I said, trying to draw him out. "What do you think the plan should look like?"

"My boss wants to know how we're going to make this happen. What are the steps? Where will we be by the end of the first quarter? How can we ensure that we're done on time and on budget?"

These were perfectly reasonable questions, if changing an organizational OS were like remodeling an office. But as you know, changing a complex system is not like changing a complicated one. There are no Gantt charts in OS transformation (unless you're okay with a lot of rows

that say "try things, learn, iterate"). This realization would not sink in at Control Inc. for a long time.

"We have a plan ... but it's written in pencil," I replied. "We're going to keep inviting teams to take ownership of their way of working. If they're not the ones naming the problems holding them back, and they're not the ones suggesting solutions, then you're not evolving your operating system. We're going to ask you and other leaders to create and hold space for experimentation. That means making it safe to spend time on this work and safe to fail. I have to tell you, so far we're not doing great on that front. People are not showing up."

"I think that's because everyone is wondering what *it* is," he interjected.

"Ah, yes, *it*," I said. "You want a destination, but what we're actually doing is starting a pattern of continuous improvement—a habit of getting better every day. You want a more capable, adaptive, and fulfilled culture. So we have to figure out what to add, in terms of new principles and practices, and what to take away, in terms of organizational debt and assumptions, that will move us in that direction. That's the 'it' you're looking for. And that needs to happen in your teams, not as an edict from the boardroom."

I paused to make sure he was still with me.

He looked at me skeptically. "Yeah, *I* get it, but I can't go back to the leadership team without a plan." And round and round we went.

Emergent Inc.

Meanwhile, a company of similar size—let's call it Emergent Inc.—was approaching transformation in a different way. It too had been confronted by a pattern of slowing growth, but more than that, its people had become frustrated with the bureaucracy and constant negotiation

inherent in their culture. Some of the COO's first words to us were "There has to be a better way!" By the time we met, she had been reading about new ways of working and thinking about specific opportunities within her culture. Could she and her colleagues make faster *and* better decisions? Could they use transparency to coordinate efforts more effectively?

Rather than weeks of debate about a plan, our project kicked off with a multiday off-site that included the top forty leaders in the company. We gathered in a circle and questioned their operating system—how it was or wasn't serving them. We heard and told stories. But we didn't just talk; we *experienced* new ways of working through activities and exercises designed to challenge assumptions, and we talked about what we noticed. We moved as much as we sat still. We heard from everyone in equal measure, asking louder voices to step back intermittently so that we could draw out the quieter voices in the room. We asked everyone to model vulnerability and candor. Seventy-two hours later, we had transformed a dozen tensions in their OS into safe-to-try experiments that small teams had committed to testing immediately. No one left wondering what was next. They left with things to do. They left with commitments to one another.

A couple months later, the COO expressed a concern of her own. "We're doing all these little experiments around how we meet and decide and share, but when do we address the big stuff, like structure or the budget?"

"Well," I replied, "the things we're doing now reflect what the teams are ready to do—what they identified as holding them back. As they master these things, they'll naturally gravitate toward bigger systemic opportunities. When teams start calling for more dynamic budgeting, you'll know exactly where to try it."

"I love that," she agreed. "But people keep coming to me with interpersonal problems that they believe are caused by our structure. I'm

wondering if I should just make those changes now and solve their tension."

"What would happen if you stopped playing the role of fixer?" one of my colleagues asked.

"I don't know . . . ," she said. "I guess they would have to figure it out for themselves?" We all looked at one another and smiled.

Two Transformations

Control Inc. and Emergent Inc. don't exist. But that doesn't mean the two stories I just shared aren't real. They happened. In fact, they've happened more than once. Over the years we've taken part in dozens and dozens of change efforts. Some worked. Some didn't. And many were somewhere in between. Control Inc. and Emergent Inc. represent amalgamations of these experiences. Emergent Inc. draws from efforts that largely succeeded in their quest to unlock the potential of their organization, sometimes far beyond our expectations and often in surprising ways that we could never have anticipated. Control Inc. draws from the many wrong turns and outright failures that we've witnessed and played a part in. Their stories continue as we move forward.

THE FOLLY OF THE CHANGE PLAN

In the popular approach to change, the one you know and have probably experienced, we imagine the future we want and then we attempt to close the gap between how things are and how they should be. Because we are oriented around a specific vision of the future, we tend to view change as finite. It's a journey from point A to point B, with a beginning, middle, and end. Once we achieve that goal, we are *done*. To that end, we manage change in a linear fashion, with a project plan, time line, milestones, and all the trappings of a controlled and ordered process.

This is a compelling but ultimately incoherent theory of change. It's compelling because it reflects what we want to believe about the world and our power to change it. Don't forget, we're carrying a hundred years of Taylorism around with us and the wiring is deep. Our entire career, we have been trained and drilled on the importance of planning, control, and finding the one best way. We want to do change *by the book*.

The most prevalent change framework we encounter, and there are dozens, is almost certainly John Kotter's eight-step process, from his 1996 best-selling book *Leading Change*. In his research, Kotter compared failed transformations with successful ones and found that eight specific actions spelled the difference. I guarantee you've heard them all before: establish a sense of urgency (what some people call a "burning platform"), build a guiding coalition, develop a shared vision, and so on. In the decades that have followed, these steps have become a kind of change dogma, wielded from the top by executives and consultants who do them *to* (rather than with) their organizations. Rather than hold these "steps" lightly, executives tend to view them as best practice and attempt to implement them linearly and authoritatively, often with lackluster results. Kotter himself updated the process in 2014, acknowledging that the steps should really be done concurrently and continuously.

The problem isn't that we use models and frameworks to better understand change (although we need to be careful about correlation versus causation when it comes to defining what "works"). The problem is that we mistake the organization for an ordered system. And so we oversimplify. As a result, we tend to force whatever is happening in the system—in the hearts and minds of our colleagues—into that framework. We start saying things such as "You're frozen right now, Michael, and we're in the 'unfreeze' part of our change process, so . . . we're going to need you to change your attitude." We can guess what Michael is thinking in response.

This plan-driven approach leaves little room for adapting as we go. After all, what happens if we learn something that fundamentally changes our vision (you know, the one we already communicated to the whole company)? Do we start over? At a scale of ten thousand or a hundred thousand people, can we even truly be in one phase together at the same time? Or are we actually experiencing different contexts and realities simultaneously? Aren't we more likely to experience change locally and asynchronously—in many different places at many different stages? The answer to these questions is simply incompatible with traditional change management.

The truth is that in an increasingly complex world, plans are lies committed to paper. They're mere horoscopes drafted by the people furthest from the action that diverge from reality the moment they're published. Change plans are no exception. In fact, they're worse. Getting everyone to execute in service of a strategic plan using their existing habits and practices is one thing. Getting them to change those habits and practices (and the mindsets that support them) is something else entirely.

Perhaps this is why so few transformations succeed at all. A recent McKinsey study showed that only 26 percent of transformation efforts succeed in the eyes of the people involved. If you ask frontline employees only, that drops to just 6 percent. Based on those statistics, you're almost as likely to hit blackjack on your first hand at a card table as you are to complete a transformation that is felt and recognized by the front line. Clearly, something has to give.

CHANGING HOW WE CHANGE

We need a new approach to organizational transformation. We can start by accepting that organizations are complex adaptive systems, not complicated mechanical ones. They are *living* systems, not machines. They

are the sum total of the principles, practices, mindsets, assumptions, and behavior of hundreds or thousands of people.

How do you change a complex adaptive system? *You live in the now.*

A city planner in search of more green space in Manhattan could envision an idyllic future where a new city park is built out into the Hudson River. But her chances of realizing that vision would be slim to none. Even armed with support and funding, construction could take *years*. The Second Avenue Subway was conceived in 1919. It didn't open until 2017. Complexity will surprise you.

Alternatively, our planner could look up and notice that an abandoned elevated railway running from the Meatpacking District to Chelsea has been colonized by wild plants, creating what looks like a linear park in the sky. That's exactly what Joshua David and Robert Hammond noticed before they founded Friends of the High Line, a nonprofit dedicated to the railway's preservation and reuse as a public park. Upon opening, the High Line was an instant phenomenon, and today it attracts more than five million visitors a year. In an unexpected twist, the buildings and real estate around the park, previously an eyesore, have now skyrocketed in value. The areas along the High Line are among the most desirable in the city. All because someone started with what was (almost) there.

The history of human innovation is a story of happy accidents. Airbnb didn't happen because of a vision to change travel and hospitality forever. It happened because a couple guys with an air mattress were short on rent. Their idea worked, so they did more of it. As leaders, we have been told we have to imagine a compelling future and drive everyone toward it. While this can gather people to our cause, it doesn't do much to further it.

Complexity expert Dave Snowden offers enigmatic but essential advice on the matter. "Managing the present to actually create a new direction of travel is more important than creating false expectations

about how things could be in the future." What he's getting at is the difference between *closing the gap*—trying to achieve a predetermined future state—and discovering what author Steven Johnson calls *the adjacent possible*. In his words, "The adjacent possible is a kind of shadow future, hovering on the edges of the present state of things, a map of all the ways in which the present can reinvent itself."

As leaders, we have to accept that we can't control whole communities of people—we can nudge them only in directions they are disposed to go. When attempting culture change in complexity, Dave Snowden says, "we manage the emergence of beneficial coherence within attractors, within boundaries." That's a mouthful, I know. In the common tongue it might read something like this: we try new things, notice positive and negative patterns, amplify what's working, and minimize what isn't. Sounds an awful lot like the process of evolution, right? If you need an approach to change, the engine of all life on this planet is a good place to start.

> The greatest loss of time is delay and expectation, which depend upon the future. We let go the present, which we have in our power, and look forward to that which depends upon chance, and so relinquish a certainty for an uncertainty.
>
> —Seneca

What is counterintuitive about this approach is that it has the potential to go much faster than the traditional one. While it may feel fast to announce a global org change (complete with org charts), when new behavior doesn't follow, reality will soon sink in. And the conversation will turn to how people resist change. But do they? Or do they just resist incoherent change efforts?

In a system that's not overly constrained, changes that serve us can spread quickly. Niels Pflaeging holds a similar view. He contends that "milk in coffee is a more helpful metaphor than the widespread notion of seeing change as a journey from here to there." Like milk in coffee, the

right kind of change can unfold throughout a system quickly and continuously.

At this point you might be realizing, Oh shit, we're not just changing the organization; we're changing *how we change the organization*. You're right. And it's every bit as meta as you're imagining. By accepting that your organization is complex, you are compelled to change both the OS *and* the manner in which you change it. You can't blow up bureaucracy with a bureaucratic change process. You can't build a culture of trust with a program full of oversight and verification. Start the way you mean to finish.

While we can't pick a specific future state and realize it through sheer force of will, that doesn't mean we can't steer in the direction of our values. We want to create organizations that are People Positive and Complexity Conscious—full of humanity, vitality, and adaptivity. That means we should measure the results of our experiments, probes, nudges, and flips according to those ideals. Will a change we make increase our ability to respond to changing circumstances? Improve our relationships or social density? Bring more meaning to the workplace? Or will it move us away from those things?

As an example, inspired by the stories I shared in the domain of compensation, we released salary data inside The Ready a couple of years ago. We deemed it safe to try because we had everyone's consent, and because information symmetry is a People Positive trait (transparency reduces bias and increases fairness). Then we waited for patterns to emerge. If someone got upset because they thought their colleague was overpaid, would that mean the experiment was a failure? Not at all. Our goal isn't harmony. Our goal is an ever better organization. By revealing a disparity between expectation and pay, we created a learning opportunity. Either the system is right, and the individual can learn more about how and why we value certain skills, or the individual is right and the system can be adjusted. Either path increases knowledge, trust, and fair-

ness. In the short run, it's easier keeping everyone in the dark. In the long run, a little light goes a long way.

We have a name for what happens when everyone in an organization is engaged and empowered to shape and reshape its operating system. We call it *continuous participatory change*. *Continuous* because we need to break the habit of treating change as a rare and hallowed thing. The inertia of the status quo has made change a last resort. We can make small local changes routine and find that progress compounds. And *participatory* because we need to break the habit of centralized, top-down transformation. We can distribute authority and encourage everyone to steer the organization in service of its purpose.

Every OS transformation is ultimately about achieving that state at scale. Because no matter how brilliant your principles and practices may be, the world will turn, and the organization will need to respond.

CONTINUOUS PARTICIPATORY CHANGE

Everybody always wants to know, "What does a typical OS transformation look like?" The truth is that no two look the same. In 2013 Pixar was riding high on the success of fourteen number-one films in a row. But early members of the leadership team felt that the culture had lost something along the way. In his book *Creativity, Inc.*, cofounder and president Ed Catmull observes, "One of the central tenets of our culture—good ideas can come from anywhere, so everyone must feel empowered to speak up—was faltering." In the wake of cofounder John Lasseter's #metoo-related termination, it's clear this was an understatement. Many employees were so respectful of the studio's history and track record that they questioned whether they had a right to speak up about anything. In the midst of that, production costs had been rising for years and it was imperative the studio figure out how to streamline and simplify its development process. When the firm's leadership met to discuss

how to achieve that goal, Guido Quaroni, the vice president of the Tools Department, suggested, "Let's ask Pixar's people—all of them—for ideas about how to do it." Everyone immediately got on board.

They decided to close Pixar for what would be known as "Notes Day," a cheeky reference to the feedback they regularly exchanged around films in development. No meetings. No visitors. Just 1,059 employees focused from dawn to dusk on how to improve their business. The newly formed Notes Day Working Group collected suggestions from the culture that culminated in more than 106 topics and 171 sessions. They created "exit forms" for each session that attendees could use to record proposals, next actions, and who would champion these ideas going forward. Employees were able to choose the sessions they felt inspired to attend, with options ranging from "Helping Directors Understand Costs in Story" to "Returning to a 'Good Ideas Come from Anywhere' Culture."

The energy throughout the day was electric, even for Pixar. Proposals were abundant, and dozens of ideas were implemented by teams in the weeks and months following the session. Catmull recalls, "I don't know if a metric exists to measure the impact of Notes Day, but from where I was standing, it was huge." The next day, he received hundreds of emails from employees expressing appreciation and gratitude for what the event had meant to them. One message on a Notes Day bulletin board simply read, "Do this again next year."

But why wait? Why unleash that level of participation, candor, catharsis, and innovation only once every 365 days? If participatory change is so powerful, why not make it *continuous*? The knee-jerk response is "We don't have time." But I think the more honest answer is "We don't know *how* to execute and learn at the same time." Getting an organization to the point of Pixar, where they're ready for notes, is hard enough. Keeping it going is something else entirely. That's one of the reasons continuous participatory change is so rare in mainstream organizations—a lack of willingness to figure it out. But if we stop looking for a proven process

and start looking for patterns, we can find ways to make change an everyday occurrence. In our work, we've noticed six such patterns that, held lightly, are worth encouraging:

Commitment: *When those with power or influence commit to moving beyond bureaucracy.*

Boundaries: *When a liminal space is created and protected.*

Priming: *When the invitation to think and work differently is offered.*

Looping: *When change is decentralized and self-management begins.*

Criticality: *When the system has tipped and there's no going back.*

Continuity: *When continuous participatory change is a way of life, and the organization is contributing to the broader community of practice.*

Do not mistake these for steps. They're more akin to thresholds. The order can change. They can be revisited and renewed. They can interact. But when they happen, they help. If you squint, they feel almost like a plan . . . written in pencil.

We'll explore each of them now, followed by some critical concepts that will help you navigate these moments. By the time we're done, you'll be ready to take the leap.

Commitment

If you look at the stories of cultures that have transcended bureaucracy, you can almost always trace it back to one or two individuals who saw an opportunity to do things differently. Fed up with the old way of running an organization, and often in a moment of crisis or rebirth, they begin to ask deeper questions. They start to suspect—and then to believe—that a better way of working is out there. These catalysts are often founders or

CEOs, but they don't have to be. In fact, if you're reading this, there's a good chance you are one. You're about to start something, something deep and emergent.

If you've ever seen a murmuration—thousands of starlings engaged in aerial ballet—you know how incredible emergence can be. They fly as one superorganism, turning and diving, fanning out and changing shape, from an oval to an hourglass and back again. This phenomenon has been the subject of speculation for centuries. The ancient Romans believed the Gods were communicating with them through the movement of the birds. In the early twentieth century, it was believed that the birds were telepathic. It wasn't until the age of computer science that simulations started to unlock their secrets. When one starling senses danger from an encroaching predator, its movement sets off a chain reaction within the flock. But how? Just three simple rules: avoidance, alignment, and attraction. Avoid crowding the birds around you, steer toward their average heading, and try to stay equidistant from your neighbors.

Our organizations are vastly more complex, and yet in order to succeed, we have to recognize that human nature and complexity call for a similar approach. In order to maximize our chances of success, the people in power have to commit to a few basic principles—simple rules for the elaborate ballet of continuous participatory change. I recognize that the principles below might seem radical or even impossible now. Use them as beacons if you're not ready or able to use them as bedrock. But if you want humanity, vitality, and adaptivity more than you want control, and you have the power, this is how you start strong.

Autonomy. All members and teams should be self-managing and self-organizing. Members have the freedom and responsibility to use their skills, judgment, and feedback loops to steer and serve the organization's purpose.

Consent. All policy decisions—agreements, rules, roles, structures, and resources—should be made through the informed consent of those impacted by that decision. In the spirit of agility, members may consent to using other forms of decision making, including distributing authority to specific roles, teams, or elected representatives.

Transparency. All information should be made available and accessible to all members. Individuals and teams should "default to open" when it comes to sharing data, information, knowledge, and insights.

Once you have the commitment of the people who hold power, you need to find other catalysts—to challenge you, encourage you, and learn alongside you. If you're a startup, that means recruiting people who are not just skilled but ready to reinvent how they work. If you're an established organization, that means finding the rebels in the system who are already hacking the bureaucracy.

The team you're forming is the seed of a community of practice that will eventually include everyone. They are the wayfinders who will be the first to try new things and the first to share what works. They will manage the present and nudge the system in the direction of more People Positive and Complexity Conscious practice.

"What exactly are we looking for?" our sponsor at Emergent Inc. asked me, as we started assembling a network of catalysts that could act as a launch team. "We're looking for insurgents with influence," I said. "People who have a dream and a team."

I continued, "Who do people confide in? Who do people go to when they have a good idea? Who do people go to when they need to break the rules? Who is already doing things differently?"

He sat back and thought for a second. "Ah, I can think of a few leaders who would be great for this. How many do we need?"

"Only true zealots," I said. "We're asking people to take on a second job here—to be stewards of an emerging OS. Quality over quantity."

Boundaries

Early in the transformation effort with Emergent Inc., we drew a circle around marketing and merchandising—two functions with common goals but very different mindsets and ways of working. If we could change how these teams organized and interacted, it would radiate out into the rest of the business. This level of focus allowed us to move quickly without distraction or scrutiny from the rest of the system. Only when others became interested in what we were doing did we expand the scope of our efforts and start scaling what we had learned.

All living systems have boundaries. There are boundaries of identity—such as who is (and isn't) part of an organization or a team. And there are boundaries of behavior. If you break certain rules as a lawyer, you will be disbarred and cease to be a member of the legal community. In OS transformation we need a boundary to create what author Dave Gray calls "liminal space." "Liminal" means "threshold" or "portal." Gray calls it the "in-between space that defines two things, while at the same time being neither one nor the other." He continues, "Change happens at the boundaries of things: the boundary between the known and unknown, the familiar and the different, between the old way and new way, the past and the future."

One of the challenging things about the transformation effort at Control Inc. was that we never really identified a liminal space. "Where can we start?" we asked, after several weeks of talking about the kinds of changes that might ensue. Our sponsor started rattling off cities—locations that might be open to the possibility of doing things differently. "Those all sound great to us. Let's just pick one and get started," we said, having long ago given up on the idea that there was a perfect place to begin. Any willing team was as good as any other. "We can't start chang-

ing things without the CEO's consent," they replied shakily. "And he's going to want to know exactly what you're going to do." Months later, we still hadn't found our safe space. Every possibility had a reason it wouldn't work. Not only did we have a boundary problem, but we had a commitment issue as well, a lesson we would learn more than once.

OS transformation needs liminal space to survive. We need a place in the organization where we can say: here we're going to do things differently, here it is safe to try. The important thing is that the space be protected—from the rest of the organization and the outside world. Antibodies and muscle memory will emerge and try to maintain the status quo. In the early days that space can be as small as one meeting, one room, or one team. In the end, the boundary of the firm won't even contain the space you've created.

When working with clients, we tend to start with a boundary around a team, then two or three, and then what General Stanley McChrystal calls a "team of teams"—a network of teams bound by a common purpose. I contend that this is the ideal scale for transformation. It seems no accident that it correlates with Dunbar's number—a commonly quoted cognitive limit on the number of people we can maintain stable social relationships with (roughly 150). In a startup this might include everybody. In a larger organization this often means a P&L, a location, a geography, or a function. It *is* possible to go all in and make your entire organization—even a large one—a liminal space. It just requires a greater degree of commitment and dialogue from everyone up front. Ultimately it is incumbent upon us to identify a liminal space and ensure that everyone inside it is a volunteer who feels a sense of safety and autonomy in the adventure ahead.

Priming

With a team of catalysts gathered and a boundary identified, the vast majority of change initiatives proceed with a process of discovery and

diagnosis. The story we tell ourselves goes something like this: first we'll study the culture, then we'll figure out what is broken, and then we'll fix it. It's a great story. Unfortunately, it's total fiction.

For starters, the vast majority of diagnostic tools and ethnographic approaches are flawed. Issues of bias, influence, self-report, gaming, and manipulation are all prevalent. Beyond that, the desire for high-level diagnosis leads to lowest-common-denominator insights that are hard to act upon. The survey says the organization has "low trust." Okay. Now what are you going to do about it?

Starting with system-wide diagnosis also perpetuates the belief that the organization is a complicated and autocratic system—something that can be analyzed and then fixed from above. It puts the power of interpretation into the hands of the few. "We will design the survey, we will interpret the data, and we will decide what to do." This is the command-and-control theater of change that we want to avoid.

Worst of all, discovery and diagnosis in a large bureaucratic system are painfully slow. In the time it takes to design a survey, get it approved, send it out to your employees, nag them to fill it out, get the results back, average them out, and decide on your top three objectives . . . we can help dozens and dozens of teams explore and experience their own adjacent possibilities.

Instead of diagnosis, we typically proceed with a process of experiential learning and dialogue we call *priming*. For catalysts, leaders, and teams who are willing, we want to challenge the basic assumptions we all hold about organizations and how we work in them. Most of us have not had the time to think about how we work, and why we work that way, in a long time (maybe ever). We're on autopilot. And while we might be frustrated by the systems we serve, we rarely stop to think about how we would change them if we could. Even executives with enormous power end up moving boxes and lines around rather than attempting real OS change. The idea behind priming is to pull people out of their patterns

and back onto a learning edge through play, reflection, and debate. To realize what it means to be People Positive and Complexity Conscious, we have to experience them.

We maintain a tool kit of exercises designed to help us create these moments of epiphany. One of our favorites is a game called Ballpoint. Originally created by Boris Gloger to deliver lessons about workflow, we use it today with executive teams around the world to help them experience self-organization. In the game, a team of twelve to twenty-five individuals is handed a bucket of ping-pong balls. A point is earned when a ball leaves the bucket, is touched by everyone on the team, and returns to the bucket. There are some rules: balls cannot touch one another, the bucket can't be moved during play, balls must have "air time" between teammates, balls that touch the floor are errors, etc. You get the picture. Teams are allowed two minutes of play, followed by ninety seconds of discussion and retrospection to change their approach. As each round begins, they estimate how they'll score. After every round, we tally up points and errors.

To say that teams get engaged in trying to maximize their score is an understatement. There is cheering. Hooting. Hollering. Sweating. The rolling up of sleeves. But something far more interesting than team building is under way.

As the leadership team at Emergent Inc. finished their fifth round of play, we gathered everyone into a circle for a debrief and conversation. "What did you notice?" we asked. It was quiet for a moment, and then the observations started flowing. "We had to change our approach several times." "We needed to communicate more." "The people on one end weren't being heard by the people on the other end." "We didn't freak out about errors; we just moved on." "Good ideas came from multiple people in the group." "There's no way we would have figured that out by just sitting there and planning." "The reflection time was key." "Performance was relative; whatever we did last time, we just wanted to beat that."

"And who was the leader?" I wondered out loud.

"Nobody."

"We all were."

Many other lessons were revealed through the game, but I won't spoil your ability to experience it by sharing them here. Suffice it to say that many of the things they needed to realize about their OS were laid bare in just forty-five minutes of play and discussion. We finished by comparing how they played as a team with how they worked day to day. Then I offered my final provocation: "You do realize that you just had more fun removing and replacing ping-pong balls from a bucket than you do on a regular day at work." They laughed. "How we work matters, doesn't it?"

If you're willing to search, there are games and activities for learning or unlearning almost anything. Enemy/Defender, an improv game, shows us how simple rules can create complex behavior. The Marshmallow Challenge highlights the importance of testing and learning. The Cynefin Lego Game illustrates, in a hands-on way, the difference between simple, complicated, and complex systems. An Identity Walk can shine a light on diversity and privilege in a way that cuts deeper than any presentation on inclusion or unconscious bias. There are more good options than you'll ever have time for.

Typical topics we like to prime include complexity, emergence, self-organization, organizational debt, agility, leanness, motivation, self-awareness, mastery, trust, generative difference, psychological safety, and more. It's the twenty-first-century curriculum that's never taught in business school. The more time we have, the more breadth and depth we can cover. But there's a risk in diving too deep and missing the chance to make practical connections as we start the work of change. We aim to do just enough priming to get hearts and minds open, always linking things back to the present moment and channeling that temporary energy into change. We can always go back for more learning (or unlearning) when the need arises.

Your goal is to prime every team inside the liminal space. It doesn't have to happen all at once. It doesn't even have to happen before they start changing. But it helps. A few hours of play and discussion sets everyone up for success in things to come. At the end of every priming experience, you're extending an invitation, not a mandate: Join us. Try something. Start a conversation. Ask for help. We are here for you. A whole new way of working is possible. Step into the space.

Looping

Earlier we talked about designing change to reveal the adjacent possible. Within organizations, we endeavor to do this through a process we call *looping*, inspired in part by Jason Little's *Lean Change Management*. In our work, a loop contains three stages that are practiced recursively: Sensing Tensions, Proposing Practices, and Conducting Experiments.

The Looping Process

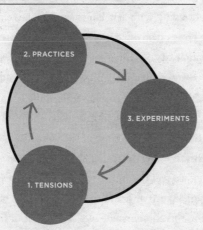

Looping is at the heart of continuous participatory change. Loops can happen at different speeds, at different times, at different scales, in different parts of the organization. At scale, this is nothing more than a distributed pattern of experimentation. But what it can lead to is truly remarkable. While we're often asked to teach clients new ways to decide, coordinate, and grow, the only thing they really need to know is how to loop. If you can loop you can learn, and if you can learn you'll figure out the rest. So let's take a deeper look at this process and how to bring it to life.

1. Sensing Tensions

I first heard the term "tension" used in an organizational context in Peter Senge's book *The Fifth Discipline*. In it he briefly introduced a concept called *creative tension*. "Imagine a rubber band, stretched between your vision and current reality. When stretched, the rubber band creates tension. . . . What does tension seek? Resolution or release." In Senge's view, creative tension must be resolved by either pulling reality toward the vision (change) or pulling your vision toward reality (compromise). If you feel, for example, that your organization has an insufficient parental leave policy, that's a tension. You can try to change it, or you can accept it. Simply living with tension is unproductive. While we don't want to get too hung up on our visions for the future (remember, we are operating in complexity), there is something deeply resonant about the idea that we can sense potential in our own systems and that we need to tune into that ability and put it to use.

Twenty-five years later, the term tension surfaced in an organizational context once again, this time in the context of Holacracy. In his book, founder Brian Robertson talks about tension as a signal that can guide us. "When we feel that sense of frustration at a system that's not working, or a mistake that keeps getting repeated, or a process that seems inefficient and cumbersome, we are tuning in to a gap between how things *are* and how they *could be*."

Hearing the word "tension" may trigger anxiety, negativity, or stress, but the concept is not inherently positive or negative. It is simply about potential. Call it an idea. Call it a spark. Call it a challenge. The important thing is that we are sensing it all the time. And no matter how mature or sophisticated a culture becomes, tension is always present in every living system. Without it we cannot grow. Without tension we are dead.

Working with teams around the world, we hear about their tensions every day. In fact, we ask for them. "What's stopping you from doing the best work of your life?" is the first question we ask at the outset of any loop. "What's stopping the organization from achieving its purpose?" is another. The range of answers is simply stunning. We have collected hundreds and hundreds of tensions over the years. While the full breadth is vast, we have curated a list of seventy-eight tensions that cut across all twelve dimensions of the OS Canvas and provide provocative fodder for incredible conversations. Take a look and see how many apply to your team or organization.

❑ Lack of trust.

❑ It isn't always clear who has the authority to make decisions.

❑ Bottlenecks in decision making.

❑ Too much consensus.

❑ People have to ask for permission to take action.

❑ Opinions matter more than data.

❑ We have meetings to prepare for meetings.

❑ Our meetings are theater.

❑ Meetings don't lead to decisions or action.

❑ No time in the day for actual work.

❑ Only the loudest voices get heard.

❑ Too many meetings.

❑ Not enough transparency.

❑ Information is shared only on a need-to-know basis.

❑ Lack of honesty and candor.

❑ Lack of visibility between teams.

❑ Too much email.

❑ The "why" behind decisions isn't always clear.

❑ Outdated tech tools.

- ❏ Our technology and tools are not maximizing our potential.

- ❏ Our processes get in the way of the work.

- ❏ We operate like everything is a crisis.

- ❏ We are addicted to planning.

- ❏ Work/life boundaries are not used or respected.

- ❏ We reflect infrequently.

- ❏ We plan and predict rather than test and learn.

- ❏ We change slower than the outside world.

- ❏ Rigid structures prevent experimentation.

- ❏ Failure is unacceptable.

- ❏ Good ideas get lost in bureaucracy.

- ❏ We don't take risks.

- ❏ Lack of feedback.

- ❏ Learned helplessness.

- ❏ Our people don't have the skills we need.

- ❏ We aren't self-aware.

- ❏ Career paths are unclear.

- ❏ We work in silos.

- ❏ Cross-functional relationships are strained.

- ❏ Roles aren't clear.

- ❏ Everyone needs to be involved in all things.

- ❏ Teams are not cross-functional or cross-disciplinary.

- ❏ Our structure inhibits collaboration.

- ❏ Too many layers.

- ❏ Misalignment between our strategy and our structure.

- ❏ No alignment on strategic priorities.

- ❏ Priorities are unclear.

- ❏ We waste time on the wrong things.

- ❏ We spend too much time trying to predict the future.

- ❏ Strategy is driven from the top rather than from the edge.

- ❏ We don't know how to say no.

- ❏ Limited resources.

- ❏ We don't leverage talent across our system.

- ❏ We drive constantly.

- ❏ The budgeting process is slow and unresponsive.

- ❏ Politics and turf wars.

- ❏ We are a culture of who can get closest to the leader.

- ❏ Individuals over teams.

- ❏ People talk behind backs.

❑ Lack of accountability and ownership.

❑ People don't do what they say they're going to do.

❑ Lack of vulnerability.

❑ Our "professionalism" gets in the way of being authentic.

❑ We favor executing over relating.

❑ We are unable to attract and retain top talent.

❑ We avoid tough conversations.

❑ Gender balance is off.

❑ Diversity isn't valued.

❑ Passive-aggressive culture.

❑ Our purpose is unclear.

❑ Not enough recognition or reward.

❑ People don't feel that their work is meaningful.

❑ We recognize and reward individuals instead of teams.

❑ We are a culture of fear.

❑ Lots of effort goes into protecting and stroking egos.

❑ We marvel at our own complexity.

❑ Shareholders are prioritized over customers and employees.

❑ We prioritize the short term over the long term.

❑ We use good (or bad) results as an excuse not to change.

That's just a sample of what real teams—teams inside the most influential institutions in the world—are struggling with. Your team has tension inside it too, just waiting to get out. It is your organization's potential, its future, its life force. Teams that understand this approach view looping as an exercise in going from good to great. They see it as continuous improvement, rather than triage or life support. Others see the organization as an ordered system that is *broken* and must be *fixed*. To their chagrin, it never is.

For a team to start looping, they simply have to notice and name a tension. But this can be harder than it sounds. Many teams lack the safety required to even talk about their tensions out loud or in front of leaders. Beyond that, many struggle to find the words to describe what they're sensing. To make things easier, we have developed a deck of cards featuring the most common tensions that helps teams push beyond

these points of resistance. When working with a team, we'll hand them a deck containing all seventy-eight tensions and challenge them to narrow it down to just *seven*. These should be the tensions they feel are most present in the team or culture and would be the most impactful to address. What's transformative about this process is that it gives teams permission to talk about the tough stuff. The card says it, so they don't have to. There it is, on the table, staring up at them.

During our initial off-site with Emergent Inc. we had a large-enough group to send three small teams off with their own decks of tension cards. Their instructions were to choose their top seven and bring them back to the group. As I walked past each team, I noticed that each one was taking its own approach to narrowing them down—some were reading them one at a time, while others had laid them out and were moving them around based on their feelings. These kinds of activities are such a mirror for social dynamics. We can learn a lot from how they approach the task.

When the groups returned, they lined up their choices adjacent to one another. Their next task? To narrow these twenty-one down to a final seven that they all shared. Several tensions had been selected by all three teams—those automatically made the grade. Then they began to deliberate and things got a little intense. Noticing that one of the top tensions was about empowerment, the leader spoke up. "I don't know how many times I have to say this, but you all are empowered to make decisions," she barked. Everyone shuffled nervously and stared at the floor. Finally, a tenured member of the group spoke up. "Yeah, I know *you* believe that, but that's just not how it feels for us day to day." After letting that breathe, I jumped in: "It's very common for there to be a few perspectives on any issue. Our job here is not to debate whether it's true but to recognize that it's true for some of us and consider what we might do about that."

2. Proposing Practices

Once a tension is identified, we can look for ways to probe it. In some ways, this is the hardest part of the process. Given the dominance of traditional management, it's unsurprising that most teams have limited experience with alternatives from which to draw inspiration. Having never before considered new ways to make decisions, for example, how can teams be expected to suggest novel approaches on the fly?

One way to counter this is to look beyond our own walls—beyond our own industries. How are different organizations solving the problems you face? Who has tried something radical and found success? The best way to find out is to go explore. From here on out, you are a student of new ways of working and the world is your laboratory. Nothing beats going and seeing for yourself.

While the outside world is vast, there are also benefits to becoming more knitted internally. One simple approach is to set up a #ways-of-working channel on your internal messaging app (if you have one) and start sharing and discussing links that highlight novel practice. Every day stories are published that share the wisdom of new (and old) ways of working. We have to open ourselves up to these stories and the lessons they contain.

Inspiration aside, most teams find it difficult to pair tensions with practices that will generate learning rather than a grand vision of a fantasy future. It's not uncommon for a tension such as "Our processes get in the way of the work" to result in a proposal such as "Fix all our processes" during a team's first attempt at looping. It's admirable but practically insurmountable. With that in mind, we often jump-start the practices conversation by bringing some good options to the table. In addition to our tension cards, we have also developed a deck of practice cards curated from Evolutionary Organizations and our own OS. Which are you willing to try?

- ❏ Craft a clear and compelling purpose for the organization.

- ❏ Craft a clear and compelling purpose for every team and every role.

- ❏ Ask teams to share their essential intent for the next six to twenty-four months.

- ❏ Clarify the metrics that matter and use them to steer.

- ❏ Recognize and celebrate noble failure.

- ❏ Replace "Is it perfect?" with "Is it safe to try?"

- ❏ Give everyone the freedom to choose when, where, and how they work.

- ❏ Clarify the decision rights held by teams and roles.

- ❏ Use the concept of a waterline to create guardrails around team and individual autonomy.

- ❏ Start distributing authority to the edge of the organization.

- ❏ Crowdsource and eliminate policies and processes that no longer make sense.

- ❏ Start by stopping a meeting, process, or habit that is holding us back.

- ❏ Leverage decision science to reduce bias and make objectively better choices.

- ❏ Use Integrative Decision Making for important collective decisions.

- ❏ Distinguish between reversible and irreversible decisions and treat them differently.

- ❏ Define spending thresholds below which no approval or advice is necessary.

- ❏ Replace the permission process with a robust advice process.

- ❏ Trade autocracy and consensus for governance by consent.

- ❏ Form SLAM teams—self-managing, lean, audacious, multidisciplinary—around critical initiatives.

- ❏ Decentralize some of the work of central functions to teams at the edge for greater context and speed.

- ❏ Abandon singular job titles and descriptions for modular roles and role mixes.

- ❏ Develop and define the roles and accountabilities inside every team.

- ❏ Invite teams to create and edit their own roles.

- ❏ Fill leadership roles through consent or election.

- ❏ Allow people to hold multiple roles on multiple teams.

- ❏ Move from static to dynamic teaming—a marketplace of roles, teams, and projects.

❏ Create a skills database to help teams find knowledge and mastery across the org.

❏ Ensure our project and investment portfolio contains sure things and wild swings.

❏ Trade "perfect" execution for constant learning and iteration.

❏ Use even over statements to make strategic priorities and trade-offs explicit.

❏ Trade traditional planning (prediction) for scenario planning (preparation).

❏ Trade fixed performance targets for relative performance targets.

❏ Move from an annual budget to a dynamic budget.

❏ Let people vote with their feet and choose the projects they believe in and want to energize.

❏ Start each period with a zero-based budget and virtual investment to capture the wisdom of the crowd.

❏ Set aside funds every quarter to be allocated by the team using participatory budgeting.

❏ Invite everyone to spend 20 percent of their time (or more) working on whatever inspires them.

❏ Break the work into sprints to learn faster and reduce risk.

❏ Limit work in progress to a specific number of projects, initiatives, or tasks.

❏ Eliminate all status updates, project reviews, and other bureaucratic theater.

❏ Eliminate or repurpose one-on-ones that gravitate toward permission or politics.

❏ Hold regular governance meetings to update agreements, rules, policies, roles, and structures.

❏ Elect a facilitator and scribe on every team to keep meetings productive and documented.

❏ Use a meeting moratorium to rebuild our operating rhythm from scratch.

❏ Learn and leverage proven meeting structures that consistently move the work forward.

❏ Hold regular retrospectives to build learning into every team, project, and initiative.

❏ Begin meetings with a chance to get present and check in as human beings.

❏ End meetings by observing what we—as a team—can do better next time.

- ❏ Take turns or speak in rounds to hear all voices during meetings and calls.

- ❏ Give up preplanned agendas and start building them on the fly.

- ❏ Create dashboards that make team activity and performance visible.

- ❏ Make org and team financials transparent and accessible.

- ❏ Make compensation transparent to everyone in the organization.

- ❏ Adopt a policy of "open by default" when it comes to information.

- ❏ Make all available information searchable and accessible.

- ❏ Work in public by making workflow and work in progress visible to other teams.

- ❏ Stop sharing files and switch to software that supports real-time collaboration.

- ❏ Ensure that all agreements, rules, policies, roles, and structures are transparent, documented, and governable.

- ❏ Phase out internal email and move to Slack, Teams, or Workplace.

- ❏ Institute a regular ask-me-anything meeting that's open to everyone.

- ❏ Prioritize generative difference when hiring and forming teams.

- ❏ Stop hiring for culture fit and start hiring for what's missing from the culture.

- ❏ Create a team charter for every team, project, or initiative.

- ❏ Create a "user manual to me" for every member of the team.

- ❏ Make time for gratitude, recognition, and celebration.

- ❏ Replace annual performance reviews with continuous feedback.

- ❏ Share a round of instant feedback after every sprint, event, or milestone.

- ❏ Ritualize and master the practice of giving, receiving, and acting on feedback.

- ❏ Create communities of practice for knowledge sharing and development.

- ❏ Transition from individual rewards to collective rewards.

- ❏ Conduct a start, stop, continue exercise on our portfolio of projects and initiatives.

- ❏ Take the time to get to know one another.

- ❏ Make time for fun.

- ❏ Create a forum for saying what needs to be said without fear of reprisal.

- ❏ Offer peer-to-peer master classes taught by team members.

- ❏ Use videoconferencing to increase emotional intelligence during remote meetings and calls.

- ❏ Limit teams and committees to fewer than nine people.

- ❏ Develop a conflict resolution process and highlight productive conflict.

These cards remove some of the friction around imagining new possibilities and trigger questions that move us forward. While sorting through the cards, someone is going to say, "Wait, what the heck is *dynamic teaming*?" And from that moment forward, the conversation has shifted from push to pull.

That said, the occasional push can be helpful. We have found over the years that some practices are a bit like yoga poses. They can train the mind and body even before we fully understand them. One such practice is the act of "checking in" to meetings with our colleagues. While most meetings already start with casual conversation, they do not formalize the process of getting present. Asking everyone to answer a check-in question such as "What has your attention?" or "What color is your mood right now?" provides two unexpected benefits. The first is that the team has time to get present in a ritualized way. Starting differently short-circuits old patterns and puts us in a headspace to behave differently throughout. But more important, checking in by hearing from everyone sets a precedent of equity of voice that is sorely missing in most teams. Just getting those with loud voices to pass the baton and those with quiet voices to use them in a meeting is a tiny revolution.

And yet very few teams will naturally gravitate toward "checking in" as a practice to adopt out of the gate, because teams that are new to looping struggle to realize just how powerful a small change can be. While many practices are indeed novel and experimental, some offer more predictable benefits if the team is willing to try a proven pose. Seemingly

> It seems that perfection
> is attained not when there
> is nothing more to add,
> but when there is nothing
> more to remove.
>
> —Antoine de Saint-Exupéry

innocuous practices that keep showing up within cultures and teams you respect are probably worth experiencing.

And finally, sometimes the best practice is no practice. In heavily bureaucratic systems, often the smartest thing you can do is remove obstacles. That might mean eliminating a policy, a layer, a meeting, a budget, or a project to see how the culture responds. So while everyone is giddy with the excitement of what we should *do*, *add*, and *try*, take the opportunity to ask them what we should stop doing. Strip the system down to a few simple rules with a healthy boundary, and you'll be amazed at what emerges (often exactly what is needed).

At this stage of the loop, every team must identify not just a range of potential practices but what they themselves are willing to try. The hurdle here is about commitment. By focusing the conversation on what we as a team are willing to try, we keep the motivation intrinsic and the stakes high. This is not, like so many corporate workouts, a theoretical discussion. This is not about choosing for someone else. This is about us.

At Emergent Inc. we considered practices from outside the firm but also from within. It turned out that two divisions in Europe had already started down a path toward a more People Positive way of working, and it was paying dividends. Both units were reporting higher revenue and retention. So we decided to go to the source. As we brought more and more teams into the looping process, we routinely invited them to check in with their progressive counterparts for ideas and learnings. The Heath brothers dubbed these existing success stories "bright spots" in their best-selling book *Made to Stick*. This approach—of studying and replicating what's already working for other teams inside your business— aligns well with our model for change. Beneficial coherence can happen anywhere. When you see it, seize it.

3. Conducting Experiments

With our tension identified and a sound practice in mind, it's time to get our hands dirty. But remember, this isn't about implementing your perfect vision; it's about living in the now. Our goal isn't to *make* something work; it's to learn, to see what's possible—what emerges. To that end, we're going to take a moment to design a thoughtful experiment.

It's easy to say, "Let's try it!" It's harder to define what that means. Who? Where? When? So while there's some value to going with the flow, we have found that teams new to continuous participatory change do better when they take a moment to answer these questions thoughtfully.

So what do we need in order to proceed? The worksheet below contains the key questions that will push a team for clarity on what they're doing and why they're doing it. We recommend individuals or teams take the time to fill this out carefully before sharing it with those who will be impacted by their proposal.

Experiment Worksheet

Tension	What is your tension? How does it manifest? Share a story that brings it to life.
Practice	What do you propose we try? What is your hypothesis? How does this practice support our commitment to a People Positive and Complexity Conscious OS?
Participants	Who will be involved? What are they committing to?
Duration	How long will the experiment last? When will you conduct a retrospective to collect perspectives and learning?

Learning	How will we know if it was beneficial or harmful?
Metrics	What kinds of stories do you hope to hear?

Requirements	What do you need in order to conduct the experiment, in terms of resources, space, supplies, funding, etc.?

Safety	What kind of support or consent do you need to make this safe to try?

With these answers in hand, an individual or team can propose their test and seek consent. Depending on the authority and autonomy present in that team, that might mean in-group consent (we as a team commit to try this) or it might mean obtaining consent from a superior or another function. Since that can be challenging, we advise using the consent process outlined on page 70. Once a team has consent to proceed, we are off to the races!

Experiments will be as varied as the tensions they aim to solve. The important thing about an experiment is that we're working on a real issue *while* working as a dynamic team. One might involve using video for conference calls, which is technologically easy but behaviorally challenging. Another might eliminate a meeting but trigger new tensions in the system—things people need that they're not getting. The task is to introduce or remove the practice clearly (and often repeatedly) and notice what happens.

Smaller, shorter practices—such as the check-in process mentioned earlier—might be easy to try and iterate over the course of a week. Larger or more complex ones—such as a new way to allocate resources—might require a little onboarding time and several weeks or months of support to see them through. Your ability to be successful in any loop is directly linked to the degree to which participants are committed to trying.

However, after reasonable exposure, be careful about forcing additional practice. While all potentially beneficial practices should get a fair shake, you also need to let behavior speak for itself. Are participants demonstrating pull—a desire for more of what you're doing? Is the practice spreading to other teams without your help? These are sure signs you are on the right track. This is where judgment comes in. When met with resistance, it can be hard to tell if a breakthrough is around the corner or you're barking up the wrong tree.

There is one very real challenge around looping that I want to make you aware of. Operating systems are interdependent and self-reinforcing. That means that an isolated experiment might not work. Not because the change was ill conceived but because it relies on a broader context to be successful. The classic example is a culture that suddenly empowers everyone to make decisions without considering the rest of the OS. What happens then? Lacking the information, tools, and sense of security required to take action, people hold back. "Nobody is stepping up and making decisions. They must not be natural leaders," management observes. But what actually happened was that they skipped ahead of the adjacent possible.

What this means is that we have to adjust our expectations to match our context and commitment. If your stakeholders are ready to move to a decentralized network of autonomous teams working transparently toward a common purpose, your loops may be more radical, progressive, and concurrent. If you're on a more incremental path, they may be focused and sequential. Sometimes we have to change small things now in order to change big things later.

Looping is an adventure in uncertainty. You're going to learn a lot more by doing than I could ever share here. Get as many repetitions under your belt as you can. Think of looping as something you have to master through deliberate practice. Start small. Start local. Be patient. And stick with it. You are starting a chain reaction that will eventually

transform your entire way of working. A more human, vital, and adaptive organization is out there, just waiting to be discovered. And once the pattern of continuous participatory change starts, it can be hard to stop.

Criticality

One of my clients spent nine months looping with a diverse group of teams, and a few leaders had emerged that were actively pushing the culture to move faster. But it was unclear just what the level of commitment really was. Were we going to go all the way? Or was incremental improvement in pockets of the organization enough? I was hiking with my team in the hills of Santa Cruz when I got a call from one of our earliest adopters there. He was hoping for a breakthrough.

"Aaron, I feel like we're stuck. How are we ever going to achieve self-sufficiency at this rate?"

"You're right. We've dipped our toes, but now we need the leadership team to jump in on transparency, autonomy, experimentation . . . all the principles and practices we've been exploring."

"If we can't figure this out, I don't know if this is the place for me anymore."

In physics, the moment of transition in which a system is halfway between one phase and another is called the critical point. Malcolm Gladwell referred to the social variation of this phenomenon using a different name: the tipping point. As it turns out, there also exists a class of systems that appear to have what is known as self-organizing criticality. Rather than needing precise conditions for the transformation to occur (the way water does to boil), these systems naturally evolve to that other state.

With a wink to that concept, we use the term criticality to refer to the moment in an OS transformation where enough people are suffi-

ciently committed to the ideals of self-management that going back to
the way things were is impractical. The operating system has undergone
a phase transition and is now something else entirely. It is, for the orga-
nization in question, quite literally the point of no return.

If a firm that has reached criticality reverses course and starts rein-
troducing bureaucracy or autocracy, a mass exodus often ensues. As you
transform your organization, this phenomenon is something to be mind-
ful of—at the team level and beyond. Curious and engaged people will
connect with this work and drink it up, and eventually it will become a
part of them. In our experience, the people and teams who reach criti-
cality are often the best and brightest, which means that you're taking a
real risk. Lose steam for the transformation, and your best people start
looking for greener pastures.

I say this not to frighten you but to clarify the importance of com-
mitment at the leadership level in work that rewires an organization's
OS. This is open-heart surgery. Proceed accordingly.

What are the signs of criticality? There are many. Two that stand
out are language and spread. Language is an interesting aspect of this
work because our vocabulary for describing culture and ways of working
is limited in popular business culture. Just as the Inuit dialect has fifty-
three words for snow, as your teams' depth of knowledge grows, so too
will their vocabulary. "Meeting" becomes too generic. What kind of
meeting? "Complaint" becomes too passive. Process that tension. Soft
words such as "safety" and "sensing" and "wholeness" start to show up.
And in many cases the culture makes up a few words and phrases of its
own. At The Ready we used to call our weekly all-hands meeting "Full
Circle," which had specific meaning to us and no one else. If you see the
language changing at scale, that signals criticality. Because language is
how we make meaning.

The second and perhaps more powerful sign is what we refer to as
spread. This is what happens when the practices you're introducing

travel beyond the boundary you've set. You hear about other groups doing things differently—borrowing tactics and methods that have worked for you. Your team gets invited to share and advise other parts of the organization, or even the whole thing. When the boundaries start to blur, it means the work is emerging under its own energy, and that means, at least in your corner of the org, things will never be the same.

Continuity

The final pattern in transformation—insofar as there is an end to this process—is what we call continuity. Following system-wide criticality, the culture has in effect taken responsibility for its own development. The operating system—previously a black box—has become facile and fungible. It is now a form of commons, owned by everyone, and everyone must maintain it.

While we often refer to organizations at this stage as self-sufficient, that doesn't mean teams don't need help from time to time. In fact, at this point we encourage teams to build that help into their structure. At Buurtzorg, the roughly one thousand self-managed teams are supported by a network of eighteen coaches. Spotify has employed a similar approach. In these and other cases, teams that are struggling have the ability to engage a coach (rather than a boss) to get unstuck. The greatest athletes in the world have coaches. Why wouldn't the greatest teams?

During the looping process we start identifying team members who are especially passionate about the practices and principles we're exploring, and we invite them to become part of a coaching network within the organization. We'll gather this group on a regular basis for training and reflection and encourage them to become a resource to their team and the teams around them. Over time these roles may become informal and reputational, or they may become a career path, as in the cases above.

Embedded coaching is just one more way to ensure that the contin-

uous participatory change process is alive and well. While it may surprise the many fans who look up to these businesses and their leadership, cultures in continuity are the first to say "We are not perfect. We are not done." In fact, it's one of the primary ways to spot an organization that has progressed this far—the acknowledgment of perpetual public beta. Part of reaching this point in a transformation or the evolution of a new organization is recognizing that change is a constant. Due to internal or external forces, conditions keep changing, and the operating system has to keep pace.

That's why a community of practice is emerging among self-managed and agile organizations. They tend to share more about what they are doing, how it's going, and what they're learning. And they tend to be curious about how others are resolving the same tensions. Openly publishing the Netflix Culture Deck was an unusual move at the time the company did so. Now it's a recruiting technique used by brands around the world. I expect that in years to come, we will see more congregations of Evolutionary Organizations, coming together to talk not about the latest trend in technology, but about far more fundamental questions of how to create an ever better future.

PSYCHOLOGICAL SAFETY

"We don't want to spend too much time on the soft stuff," a leader at Control Inc. told me before kickoff at a team off-site. "We want to get into the *real* work." To most executives that means messing with the org chart, accountabilities, or metrics. Those things feel like serious business. They support our view of organizations as machines. By focusing on these tangible objectives, we can almost forget that we are dealing with human beings.

This reaction is extremely common. Many leaders have been trained to believe that making time for emotional and relational development is

optional or even counterproductive. This is reinforced by a cultural narrative that says we should keep our personal and professional lives separate. Even the statement "be professional" is laden with expectations about putting on the corporate mask. *Just do your job. Keep your feelings to yourself and your head down.* As if we were still on a factory floor at the turn of the last century.

One of the reasons we don't like to get into the "soft stuff" is that it makes us feel vulnerable. And if there's one thing we've learned climbing the corporate ladder, it's this: don't appear weak. If you're exposed, someone else might take advantage of you. Or you'll be shamed. Or worst of all, you'll be fired. And so we've actually built up antibodies to honest human conversation at work. Sure, we might let the truth fly over drinks after hours, but in a meeting, in front of everyone else? Not a chance. This translates into an aversion to risk of any kind. If we can't even be real for fear of the consequences, what are the chances we're going to swing for the fences with our next disruptive idea?

Unfortunately, cultures that don't value and nurture openness and connection become dysfunctional and eventually toxic. Control Inc. was no exception. While some of the leadership felt that working on operational efficiency was the priority, it was obvious to us that employees had been traumatized by a revolving door of layoffs and dismissals. The pressure to deliver results in a bad year had become pervasive and suffocating, stealing nights and weekends, causing last-minute shifts in priorities and resourcing, and generally wreaking havoc on the team's mental state. Employees were constantly looking over their shoulders, waiting for someone to change the game on them. In a team check-in one Monday morning, one employee actually said: "I considered getting into a car accident this morning on my way to work, just so I'd have an excuse not to come in." There was not a drop of sarcasm in her voice. Those are the moments when you know you have triage to do. Because a system in

trauma can't transform. If we're not safe, we can't trust. If we're not safe, we can't risk. And nobody does their best work scared.

In 2012, a team within Google launched an initiative called Project Aristotle. The aim of the research was to use data from actual Googlers to determine what influences team performance. The question they hoped to finally answer: What makes a great team? Observing hundreds of teams over several years, they tracked hundreds of traits to test their correlation with team performance. What surprised them was that almost nothing seemed to be predictive. Performance seemed to be almost completely contextual. What mattered in one team didn't matter in another. And they thought they might have hit a dead end. Until they noticed that, while effective teams were different in countless ways, they all seemed to share two qualities. First, all the members of the team spent a roughly equal amount of time talking. The technical term for this is equality in distribution of conversational turn taking. Second, the members of the team displayed fairly good intuition about what the other members were feeling. This is referred to as high average social sensitivity. Equal talk time and emotional intelligence. It was the so-called soft stuff!

As team members on Project Aristotle wrestled with how to describe these insights and the conditions they created within a team, they came across the work of Harvard Business School professor Amy Edmondson, a pioneer in studying and defining a social context known as "psychological safety." Edmondson describes this as a "shared belief held by members of a team that the team is safe for interpersonal risk-taking." The researchers at Google realized that Edmondson's work correlated perfectly with their data. The equal talk time and emotional intelligence they were seeing were leading to the phenomenon of psychological safety. That's what made them great. Google has since tried to operationalize and scale this concept within its ranks. At that scale you can be sure

A team is not a group of people who work together. A team is a group of people who trust each other.

—Simon Sinek

this isn't easy, but the payoff is huge. Even a small improvement in overall psychological safety can lead to breakout results.

The role psychological safety plays in OS transformation work is profound. As we saw with Control Inc., if you're living in a state of constant fear, you're unlikely to take a risk on a new approach, break with protocol, speak your mind, or act outside the chain of command—all things we're going to ask people to do in this process. In other words, an organization's capacity to adapt and evolve to meet new challenges is inversely proportional to its employees' degree of psychological safety. And that means the soft stuff is often job one.

At Emergent Inc. we decided to build some psychological safety right out of the gate, using a technique called ICBD. ICBD stands for Intentions, Concerns, Borders, and Dreams. It's a conversational exercise created by Alex Jamieson and Bob Gower that boosts trust within a team by asking members to confide in one another. We asked the group to sit in a circle and presented them with a question from each of the four areas. The question for Intentions was "Why do you personally want to participate in this change?" Then we heard an answer from each participant while the rest of the group listened. Over the years, I've seen this process elicit reactions ranging from solemn silence to full-on tears. It can be deeply emotional. Teams report feeling more connected and eager to continue having honest conversations.

Of course, some people also choose to play it safe. That's to be expected. So I kept reminding this team, "You get what you give here, folks." But I couldn't do the work for them. What I could do was model the vulnerability we're looking for. So I consciously chose to say the things coaches and consultants don't say out loud. "My *intention* is to create a case study here that I can use in my book. My *concern* is that I

won't be successful in getting this group to open up. My *border* is that I leave work at 5:00 P.M. every day to spend time with my family. No late-night emails or calls with you just because we're under contract. My *dream* is that this will spread to other parts of your organization." You could see the whites of everyone's eyes. But it helped. Others stepped up too. Somebody went ahead and said, "I'm here for the money." And they meant it. Somebody else confided, "I used to have fun at work, and I want to find that version of myself again." For a group that had barely been speaking the day before, this was *progress*. Body language started to change. People leaned in. This was strong stuff!

After nearly ninety minutes of ICBD, I turned to the group and asked, "Who thinks we're ready to do the *real work* now?" I paused. "And who thinks that *was* the real work?" I let that hang in the air for a while. I could tell they weren't quite ready to say it, so I said it for them. "Sometimes the soft stuff *is* the hard stuff. You need to get comfortable with that if you're going to lead this change. Some things we can write down. And some things happen in the space between people. *Everything* is *the work*. And you can't predict where it will lead."

THE ROLE OF THE LEADER

One of the questions I hear most frequently from leaders is "What are the conditions for a successful transformation?" And the inevitable follow-up "How can I ensure this goes well?" These are incredibly important questions. Because your role in changing your OS is different from any change you've led before. This is not the time to bang the drum and march your team into the future as you always have. OS transformation is about distributing power, and while you might want to do that with one grand gesture, it actually happens one interaction at a time.

The biggest barrier to change, believe it or not, is you. If you're the founder, the CEO, or the team leader, you hold a disproportionate amount

of power in your organization. People look to you for many things. For decisions. For permission. For attention. For feedback. For signs of what's to come. If they look at you and see the same behavior, they're going to dismiss this as another fad. That means you have work to do.

> Man has no greater enemy than himself.
>
> —Petrarch

Work to master your ego. Work to quiet your voice. Work to step out of the way. You must become a paragon of trying new things. Starting new loops. Asking big questions. Don't stay stuck in the habit of evaluating and judging the work of others. Go find something to *do*.

Your new job is to ensure that the conditions for change are in place, not just now but in perpetuity. While you won't be doing as much "leading" in the traditional sense, you'll be doing something far more rewarding. You'll be *creating* and *holding* the space for change.

Creating Space

We create space at the dawn of something new. Once upon a time, someone (maybe you) had an idea. When the pursuit of that idea became a real business, a beacon was lit. That beacon gathered like-minded people to the cause. Space, in this sense, is both physical and conceptual. It is brought to life by the intent (and often the means) to serve a specific purpose. This purpose is the force that binds an organization or team together.

For example, if you want to ensure that everyone in the world has access to clean water, a nonprofit called charity: water is one of just a handful of organizations that have created space for that work. When its founder, Scott Harrison, looked at the water crisis fourteen years ago, he didn't see an organization approaching it the way he wanted to—as a storyteller. So he started from scratch. charity: water would become *the* place to participate in the reinvention of philanthropy. Early innovations like being able to see where your money went in the field—knowing that a

particular village or individual was drinking clean water because of you—electrified a new generation of donors. Today more than seven million people have clean water because Scott planted a flag and made some space.

When you commit to changing your OS, you're taking on the challenge of creating a liminal space. And as we've discussed, continuous participatory change requires a boundary in which everyone can feel both safety and disruption. That space could be as small as a cross-functional team or as large as an entire firm. But drawing a circle around a group of people is just the beginning.

Creating space means ensuring that everyone on the team is committed to taking risks together and supporting one another. As wonderful as your culture may be, it's best to assume that some people in your organization don't feel safe. They don't feel safe to be themselves. They don't feel safe to tell the truth. They don't feel safe to fail. Even if you think they've got it wrong—you and your team have work to do to make it right. Ask people in the organization about the consequences of taking risks. Listen not just to what they say but to how they say it. And then begin to change the dynamic. Let people see risk taking and failure recognized and rewarded. Let them see empathy. Let them see inclusion.

Creating space also means making time for change. Because most teams simply don't have it. They are constantly being asked to one-up themselves, delivering more and more performance in the never-ending quest for growth. But they're juggling too many balls. To make up for what is effectively a dysfunctional operating system, they've had to adopt an unsustainable pace of work. Long days. Few breaks. And so most teams are one emergency away from total exhaustion.

One of the most self-destructive tendencies within teams is to get so busy that we believe there's no time to get better at how we work. One of my favorite comics of all time illustrates this beautifully. A few people are struggling mightily to pull a cart full of stones up a hill. The cart is outfitted with surprisingly *square* wheels. Another man who has

happened upon them offers an innovation: *round* wheels. "No thanks!" they say. "We are too busy." That is your team. That is almost every team I've ever worked with.

In order to do the work of change, we need to make time for it, and we need to be supported in that process. Continuous improvement is not magic; it is a discipline. It is a thing we *do*. And like all valuable things, it takes time. The average team doesn't spend even thirty minutes a week reflecting on how they work together. They are going to need your help to stop this pattern and clear a path. Because we can't learn if there's no time to think. Teams need time for retrospection. They need time for personal reflection. They need time for deep work. They need time to run experiments and find a better way. You can create these opportunities for them, or you can empower them to create them for themselves. Either way, you're going to have to hold the entire system (and its hunger for short-term results) at bay. If you were starting to worry that self-management might make you irrelevant, think again. You are a space maker.

Holding Space

Once the space for change exists, your second task begins: holding it. Many forces exist that will try to undermine an OS in transition. People who have a vested interest in the status quo will try to preserve it. People who don't understand something new will try to resist it. But the hardest part about holding space for change is that *you* can't fill it. Your organization is going to encounter so many questions: How should we restructure this team? What metrics matter? Who is right for this role? How should we allocate these resources? And you're going to feel compelled to speak up, because you're not used to letting questions linger. You're used to having all the answers. For as long as you can remember, your job has been to catch mistakes before they happen—to save the team from themselves. But in a future rich with complexity, that's no longer what's

needed. Your *new* job is to ensure ever-growing capability—a culture that gets better and more resilient every day. And that won't happen if you're playing the hero.

Holding space means making room for teams to figure things out for themselves—for failure, learning, and growth. You're asking teams to take responsibility for their own way of working. They have to find their own way in order to grow. Every time you give an order or an answer, you're either wrong and preventing them from discovering something better, or you're right and preventing them from learning. Holding space means being committed to building the muscle, scar tissue, and resilience that can come only from walking the path.

Does this mean you never interject? Never prevent catastrophe? Of course not. You, like everyone in a self-managing system, have a responsibility to speak up if you see something that is unsafe to try—something that will do irreparable harm to the organization or the people in it. The wisdom to know the difference—between the reversible and the irreversible—is what is asked of everyone in a self-managed system. But you can show them the way.

The duty to hold space never ends. You need to be sure, before you begin, that you're willing to defend your way of working from all enemies, foreign and domestic. When Jan Wallander took over Handelsbanken in 1970, he helped the distressed bank implement a decentralized approach that completely changed its future. And this approach kept evolving and growing, with staggering results, for more than forty-five years. Yet when Frank Vang-Jensen took the reins as CEO in 2015, the system was once again at risk. Vang-Jensen immediately began to centralize power back to Handelsbanken headquarters, frustrating a culture grounded in decades of independence. Sensing the need to protect the model that had served the company for so long, Chairman Pär Boman and the bank's board moved to replace Vang-Jensen after just eighteen months. In a press release related to the dismissal, Boman explained, "All

managers at Handelsbanken—particularly the branch managers—must have a very high degree of autonomy. Being the most senior manager at the bank therefore requires a special type of leadership—considerably more complex than traditional management." Ever vigilant, especially when it mattered most.

Creating and holding space is hard work. But here's the good news. As your organization and OS mature, a growing number of people will join your movement and connect their calling to its purpose. A new network of values, intent, and meaning will emerge. And an idea that once lived with you and you alone will make its home among other people. The purpose and space you've made will become community owned. It won't be yours anymore. You'll be part of something bigger than yourself.

PRINCIPLES FOR CHANGE

Practicing continuous participatory change over and over has surfaced some valuable lessons about how to do it well. They're not rules, mind you. In complexity, rules can become constraints that limit our ability to adapt. "Always" and "never" are words we try to avoid. Instead we attempt to develop principles or heuristics—guidelines that prove useful in certain situations without being overly rigid or restrictive. These somewhat counterintuitive aphorisms may help you avoid some common pitfalls. While you're living in the now, managing the present, and revealing the adjacent possible, remember these ideas. When you feel stuck, they may help you find a path forward.

Through Them, Not to Them

Organizational change is high stakes. It brings out the control freak in each of us, particularly leaders who are accountable for results. That's why the vast majority of culture change efforts are driven from the top.

While everyone sees the opportunity for change, only leaders (often the very top leaders only) have the power to do anything about it. When their need proves great enough, they spring into action and a change process ensues.

Mistaking a complex system for a complicated one, leaders try to control the change. They decide who will be in the know. They review and approve everything as if they were building a custom home. They review survey results and filter which parts managers get to see (some) and which parts everyone else gets to see (far less than that). They envision the future and then demand that the system conform to that vision. Which is too bad, because all the information about how the system needs to change in order to achieve their goals is out there, *in the system*. Change, for most people in most organizations, is effectively an announcement: effective today, things are different. But they know better.

Instead we encourage participation and ownership from the earliest days of any transformation effort. Long before anything is decided, we invite everyone who will be affected to join in. This entails some risk. Every member within the boundary we select has a seat at the table and the ability to know what is going on from day one. They are "looping without supervision." This is an intensely uncomfortable proposition to leaders. They immediately fear a variety of negative outcomes: that teams will unleash chaos, that candor and transparency will create panic or negativity, that they won't be able to control what unfolds, and of course the shadow fear—that they aren't really needed at all. They have to let these fears go. The more they trust, the faster the right change can occur.

Start Small

Inside large organizations, moving the needle requires big moves—big programs, big campaigns, and big acquisitions. As a result, many leaders dismiss small moves as inconsequential. Projects and programs often

end up with overly ambitious goals and time lines. Big wins delivered in the short term. Could anything be more challenging? Almost every idea for a new product, new process, or new tool is met with the same question: "How will that work at scale?" As if we were going to go from zero to one hundred on it. As if it were binary. And if it does seem promising at scale, then we default to a scaled implementation. Massive funding. Umpteen approvals. Project managers. Rollout plan. And on and on. In a simpler world, that might make sense. But in a complex world, things can change rapidly and unexpectedly. Focusing exclusively on scale slows us down, creates risk, and eliminates possibilities. The simple metaphor is a two-hundred-pound stone versus two hundred one-pound stones. Which is easier to move? Which is riskier to move? Which one can best leverage many hands?

Instead, start small. Involve fewer people. Change one part of a process rather than the whole thing, or change it for one group but not another. You'll be surprised how easy it is to scale something that has been validated and improved by the people who have to use it every day. When your team is taking smaller bites, there's a feeling of momentum, and the electricity of that is palpable. Other teams see it. They feel it. And they want it for themselves. None of this is to say that scale doesn't matter, just that it's not our first priority. Paul Graham, one of the cofounders of the startup incubator Y Combinator, advises his startups to "do things that don't scale." What he means is that in the early days of anything new, worrying about scale can prevent us from learning, growing, and being remarkable. Think systemically. Act locally. And let scale happen.

Learn by Doing

We are all experts in our current way of working. After years or even decades of practice, we know how to play the game, even if the game is flawed. And we have been rewarded for that expertise with position and

privilege. Is it any wonder that trying something new triggers fears of inadequacy and uncertainty? Who wants to feel like a beginner? Who wants to be uncomfortable?

When we can't be vulnerable, these feelings manifest as skepticism. We want to analyze before taking action. We want to study. We want time to think of all the angles and ways it could go wrong. We want to be sure that this is the *one best way* before we try. In corporate culture, the default question is "Is there any risk present here?" This is a defense mechanism designed to protect the status quo. Because there's always something we can point to as a reason not to try. But this is incompatible with complexity. We can't know without trying. This is akin to thinking you can improve your swing by reading *Golf Digest*. You might get some good ideas, but that doesn't mean you've mastered them.

Instead encourage teams to experience and experiment with new ways of working rather than discussing and debating them. If the time frame for the experiment is reasonable, we'll just try it without any fanfare and then reflect on how it helped (or didn't). We do this for two reasons. First, the amount of data you have *after* you try something is infinitely more than the amount you have beforehand. And second, most changes to a way of working are, in fact, safe to try. The worst-case scenario is we all look at one another and agree: let's not do that again. Even that lesson has value.

Sense and Respond

The typical mental model for leadership includes the notion that we must manifest success through sheer force of will. We commit to an objective and then we barrel ahead, oblivious to detractors or resisters who might stand in our way. As a result we often engage with the world around us through the lens of what we *want*. We try to shape it to match our wishes. This applies to the way we develop people, products, markets,

and everything else. The shadow side of this intensity is that we often forget to listen. We forget to sense what's actually going on, and we miss the signals and feedback that can help us navigate. Have you ever been in a daylong meeting where the leader in charge vigorously enforced the agenda even though it was clear the conversation needed to go somewhere else? That's shaping behavior. It happens in product development too. If you're on a team that doesn't have direct, unfiltered contact with your customers or users, you have no hope of empathy. Internally or externally, if you ever ask yourself the question "What am I missing?" the answer is "Probably a lot."

Instead we can choose to practice sensing. Sensing means noticing—with our heads and our hearts—what's really going on. We can sense what's happening in the room, what's happening in a relationship, even what's happening in our own bodies. Avoiding a colleague because you subconsciously feel judged by them is not progress. Noticing that you feel that way, naming it, and working on it (ideally with them) is sensing and responding. While we often think of feedback as something that's written in a survey, true feedback comes in many forms. It's only by closing the feedback loop among our intent, our sense of what's happening, and our actions that we can truly operate in an adaptive way. This applies at every level from the market to the miniscule. When we let go of our desire to shape everything and everyone and instead sense what is happening and what could happen, we are able to see the moves that will bring us closer to our goals, often in ways we can't anticipate.

Start by Stopping

Human beings are naturally acquisitive. We're far less gifted at getting rid of things. Just open a kitchen drawer, hall closet, or garage and you'll see what I mean. We add more and more to our collections, never

stopping to see what we can do without. Organizations are no exception. For every problem, we believe the solution is *more*. Additional employees. Another meeting. A new policy. Each one adds to the organizational debt.

At the outset, your instinct is going to be to introduce new ways of working. New meeting types. New decision tools. New structures. And while those are exciting and effective practices, you may find yourself building on an unstable foundation. If there are too many inhibitive practices already in place, it can be overwhelming to add new or contradictory ones to the mix. There simply isn't space.

Instead, ask yourself, "What can we remove, take away, or stop doing?" Consider the tension you're addressing from the perspective of what is preventing better behavior or practices from emerging. The vast majority of rules and process are created because we don't trust people to do the right thing. As we move toward an organization that can run itself, we start to value the open space that allows people to exercise judgment. Once you identify something that isn't serving you, propose to eliminate it for a period of time. Let the culture fill in the blanks. Rather than a new vacation policy, just eliminate the one you have. Tell people you trust them to take vacations responsibly. Rather than a new meeting, get rid of one that isn't serving you. See what people miss from that meeting that they can't get another way. The space that this creates will allow individuals and teams to find the adjacent possible more easily. Then, as it emerges, allow or even amplify a new approach that is working well.

Join the Resistance

People resist change. That's what most of us believe. We, the enlightened few, are willing to step boldly into the future. Anyone who defends the status quo must be wrong. These luddites, laggards, and late adopters

refuse to engage with our change programs. Can't they see that *this* change is inevitable and better for everyone? The train is leaving the station, and they better get on board. Right?

Wrong. This cliché perception of resistance leads us astray. It's a reality-distortion field that blunts our empathy and our humility. People can and do change. They just do so when it makes sense *to them*. People don't resist all change; they resist incoherent change poorly managed. Perhaps we're asking people to take risks while their incentives push them to avoid failure at all costs. Or we're asking them to trade mastery in one process or tool for confusion in a new one. These kinds of scenarios cause very real, often very rational human reactions. In response, we're dismissive. So they dig in further and shut down. And then we're stuck.

Instead of perpetuating this pattern, look at resistance as information. People are telling you something when they resist change. Your job is to find out what. Resistance is an invitation to talk, listen, and learn. That doesn't mean convincing or cajoling skeptics. Let them come to the work if and when they're ready. But while you wait and hold space for them, use their insights to ensure that the change you're driving is ever more coherent and inclusive.

SCALING CHANGE

"How do we scale this?" our sponsor at Emergent Inc. asked. We had been working in her part of the organization for a few months and there was growing excitement about the change thus far. Early positive signs had come in the form of speed and time saved. One of her top leaders summed it up by saying, "It's actually stunning the amount of work we've unblocked in the last four weeks that, frankly, has been blocked for years." At the same time, we were hearing about deeper and richer

conversations. "This leadership team has talked more in the past three months than in the past six years," another exec confessed during a retrospective.

But all that momentum was leading to a question: what about everyone else? So far we'd focused on a few hundred individuals in a business of more than twenty thousand. At this rate we'd never reach everyone—we'd all be retired first. So the question about scale was critical.

Scale is one of the most complex and contentious topics in the world of self-management. For starters, very few organizations larger than ten thousand employees can really claim to operate in a dynamic and decentralized way. There are some who doubt that it's even possible (or say it's not worth the trouble). Others claim that it can be done but that it requires a fractal approach—a federation of smaller groups working more like a marketplace than a traditional company. One thing is for sure: the unprecedented scale of human activity makes things harder than they need to be.

While it is undeniably easier to encode the principles and practices of self-management *before* an organization achieves scale, it can be done long after. What's more, we must learn how to change at scale or we risk an unnecessary amount of creative destruction in the years ahead. Refreshing large dysfunctional systems without total system failure is actually one of the great challenges of our time. Because while you can found another startup in a weekend, creating another school system or healthcare system or government is another matter entirely.

The most confusing aspect of searching for self-management at scale may simply be definitional. Is Toyota doing it? Is Google? Is Patagonia? Is Southwest Airlines? That depends on what *it* is. Certainly they're each bringing some of the ideas in this book to life, but none of them has abandoned the Legacy OS entirely. And even if they did, this way of working isn't a destination with a finish line and a winner's circle.

It's a journey of continuous improvement and vitality. Nobody's ever done, so categorizing organizations might not be the most productive use of our time.

Instead, if we can simply agree that toppling bureaucracy and replacing it with something more adaptive, vital, and human is crucial, the only question that matters is how to accelerate that process. We are left to wonder, What is theoretically possible?

The first thing to recognize is that priming and looping deep in the organization can lead to scale organically, but this rarely happens quickly or cleanly. Insurgence is messy. It's going to butt up against the antibodies of the existing OS at every turn. However, each interaction that primes new thinking or triggers a change is a ripple in the pond. Every experiment has the potential to catalyze other teams into starting loops of their own. I jokingly refer to this as "looping the loops." If the right conditions are in place, the adjacent possible will unfold.

In our work, we've noticed at least three positive patterns that emerge on the heels of grassroots priming and looping.

Confidence. When people are allowed to take ownership over even a small part of their way of working in a safe space, they open up. As loops compound, psychological safety increases, and behaviors change as a consequence. Trust grows. In fact, recent research by David DeSteno at Northeastern University shows that trust has a kind of halo effect that extends beyond the individual interaction that increased it. Trust begets trust. A team that has completed even two or three loops successfully without interference or restriction will feel dramatically different from the rest of the culture. And that confidence will spread into other teams as members cross-pollinate.

Relief. It's obvious but still powerful that when you save people time, decrease stress, increase performance, or make progress on inclusion,

it's noticeable—and measurable. When a CMO at a *Fortune* 10 company reported to his peers that he was saving fourteen hours a week based on changes his team had made, you could see the jealousy on everyone's face. There's nothing that legacy cultures love more than a measurable productivity gain, so when things do move the needle, they spread fast.

Wonder. Winning experiments reveal new possibilities. Working in new ways opens the mind to what could be. Clayton Christensen once told software entrepreneur Jason Fried, "Questions are places in your mind where answers fit. If you haven't asked the question, the answer has nowhere to go. It hits your mind and bounces right off. You have to ask the question—you have to want to know—in order to open up the space for the answer to fit." This is profoundly true in transformation work. We start small and expand as our hunger for more radical ideas grows. If you haven't even tried holding a meeting without a preplanned agenda, how are you going to be open to eliminating traditional budgets? You're not. With each loop, we take stock of the questions our colleagues are asking and meet them where they are. We know that good loops will lead to bigger and deeper questions.

We saw all three of these patterns unfold at Emergent Inc. around the same time our sponsor asked about scaling. One of their early experiments was focused on working in sprints. Project management had become lethargic and isolated, and that tension was front and center for the teams we knew. "We'll agree to kick off a project and then wait a month for the first meeting," a team leader explained. "We'll plod along keeping the work in a silo until the very last minute, and then reveal the answer to the org. By then it's too late to do anything with the feedback we get." For the experiment, a few teams agreed to try working in one-week sprints for eight weeks, sharing their work in progress with the organization every Friday, however raw it might be.

Within a few weeks, "sprint" had become part of the vocabulary in

that part of the business. Teams were reporting dramatically higher productivity, and they were finding that the feedback they got each week was far more actionable than before. We started to notice that teams who were not part of the initial experiment were sprinting on their own. In fact, the enthusiasm grew to the point that, for a while, sprints became the answer to every question. "Don't like something? Try a sprint!" was the mantra. While this was obviously not the solution to every issue, excitement like that showed us that we were on the right track.

While that kind of progress can be exciting, starting in one function or P&L and hoping for the spontaneous adoption of new ways of working in the rest of the business is unrealistic. There are better ways to proceed if we are having success locally.

One is to use that momentum to move upward from local success toward the top of the house, ideally spanning more than one function, department, or location. When you start too deep in the business, teams may find that they lack the authority to change what challenges them most. While they often have more autonomy than they think—most teams can change how they meet and communicate, for instance—they are limited by their span of control. Teams with more power can accelerate things in a way no one else can by eliminating constraints and distributing the authority to change. They can choose to experiment with—and even remove—processes, policies, structures, and tools that inhibit everyone. Looping with senior leadership teams almost always increases the distribution of power and information in a culture. And that in turn releases energy and potential that were previously trapped.

For that reason, whenever possible, we start at the very top. Even if the boundary we set for the transformation isn't firm-wide, having the support and curiosity of the top team makes everything go faster later on. Some communities of practice take a very hard line on this. The incredible corporate liberation movement in France, championed by *Freedom, Inc.* coauthor Isaac Getz, won't let an organization join its

ranks unless the CEO is fully committed to the principles of self-management. While I applaud this conviction, I'm not sure the only strategy we should employ is waiting for CEOs to come around. Actually *experiencing* new ways of working is the surest path to epiphany. We can always start by starting.

Back at Emergent Inc., that question about scaling was still lingering. "Rather than worry about how we scale this, let's worry about what might stop these practices from spreading," I said. That turned out to be a long list. Other leaders. IT policies that would blacklist our favorite tools. The urgency of quarterly goals. And a general lack of permission to try things, to name a few.

We decided to make two moves right away. First, we would take the time to do some priming work with the senior leadership team. Now that we had some early signs of life, they needed to be invited to participate in, consent to, or object to what we were doing.

Second, we had to create the conditions for sharing to occur. When you have a new approach that has become popular with a handful of teams, you want it to spread. For that to happen, teams have to feel safe, encouraged even, to notice what's working for others and to adapt it to their context. We decided to highlight emergent practices from across the org every month during all-hands meetings. The message was "Look at what your colleagues are doing. We support them. If you see something you want to try, we will get you the support you need."

While scaling a new way of working (and changing) can be daunting, don't allow yourself to be too overwhelmed by it. Criticality is fractal—tipping points can occur at any scale. A person can tip a team, a team can tip a division, and a division can tip a global corporation. As the saying goes, there are decades where nothing happens, and there are weeks where decades happen.

Epilogue

WHAT DREAMS MAY COME

The future is already here—it's just not evenly
distributed.

—William Gibson

DEEP DOWN, I THINK WE ALL HAVE A PRETTY GOOD IDEA OF
what will happen if we don't change how we work. We're watching it
unfold in slow motion right now. Massive bureaucracies lacking con-
science or purpose. Startups meant to disrupt the status quo uninten-
tionally entrenching it. Rampant inequality. Wage stagnation. Workers
displaced by technology funded with the profits from their labor. Nation-
alism. Hacked democracy. A stock market driven more by policy and
punditry than performance. And the coup de grâce, accelerating climate
change that threatens the safety and security of billions of people. All of
it the result of our mindless adherence to the ways of the past—to an
operating system that fundamentally misunderstands complexity and
human nature. This isn't the uplifting capitalism of a burgeoning econ-
omy, it's advanced capitalism at best and crony capitalism at worst.

That future doesn't work for me. Somehow we have to reconcile that
twentieth-century bureaucracy and capitalism got us *here*, but they won't

get us where we need to go unless they (and we) evolve. We are being called to lead that renaissance. You can feel it in your gut that *this* is the turning point. The future is up for grabs. And what I keep wondering is, What if we get it *right*? We have a pretty good idea what an Evolutionary Organization looks like. But what about an *evolutionary world*—one that's People Positive and Complexity Conscious at scale? That's much harder to imagine. But it's worth a try. In my future of work . . .

> Every business, nonprofit, government, and social institution practices **Continuous Participatory Change** and aspires to get better every day, not just for its customers but for all stakeholders, including the community.

> **Purpose** and **human flourishing** are the objectives that drive organizations. Growth is a result, not a goal.

> **Self-management** is the dominant organizing construct, and most individuals go to work in an environment that promotes freedom and responsibility.

> **Employee ownership** and **participation** are expected and supported. Not every business is a true cooperative, but rewards are far more distributed. Many organizations become employee- or community-owned over time as founders and investors realize sufficient returns.

> New forms of incorporation allow organizations to include **public benefit** as a counterbalance to their fiduciary duty. Making money is no longer the sole function of business.

> More companies stay **private** and resist outside investment to remain free of outside influence and pressure. Those that *need* funding increasingly get it from their community or investors with a broader definition of return. Those that *need* to go public do so via new platforms that allow them to optimize for the long term rather than the short term.

> **Diversity**, **equity**, and **inclusion** are considered fundamental to success, both because they improve performance and because they reflect our values. Organizations make trade-offs in order to excel in these areas.

> **Innovation** and **experimentation** are far more pervasive and distributed. It's not uncommon to encounter a trial or test, even in well-optimized environments.

> Due to **technological augmentation** and **reduced bureaucracy**, people are more creative and productive. As a result, organizations are smaller in terms of head count and generate more profit per head.

> **Sustainability** and **environmental stewardship** are recognized as a fundamental part of the economic equation. Carbon neutrality is table stakes. Regenerative design and operations are increasingly where the action is.

> **Regulation** is more **Complexity Conscious**—driven less by rules and compliance and more by experimentation and outcomes. Instead of "What fuel efficiency target must all automobiles meet by 2020," the question might be "What conditions or enabling constraints lead auto manufacturers to increase fuel efficiency continuously?"

> While volatility and economic swings still occur, they are less frequently the result of avoidable systemic dysfunction. Increased **resilience** and **adaptivity** in most firms and industries result in rapid sensemaking and responsiveness, mitigating the length and severity of damage.

> **Philanthropy** and **business** are no longer orthogonal. More companies look like Toms or Warby Parker than like Adidas or Ray-Ban. Founders increasingly identify with Ricardo Semler's sentiment that "if you're giving back, you took too much." Instead, they build social impact right into their business model.

> Consumers **shop locally** in order to increase the **flow of money** in their community. This in turn puts pressure on national and global businesses to source, operate, and invest locally.

> It is increasingly possible to shop **locally** *and* **online**. The two are no longer mutually exclusive, as local variations of many products exist and can be produced, shipped, and serviced near you by members of your own community.

> Public education prepares students for jobs that don't exist yet by focusing on **creativity** and **complex problem solving**. Students spend most of their time learning how to be part of a high-performing team. **Entrepreneurial skills** are prized above Ivy League admission.

> New forms of **universal basic income** are being tested for their ability to provide for our basic human needs while also encouraging us to use and share our gifts—through entrepreneurship, service, and community.

> New forms of **currency** and **means of exchange** provide alternatives to the current model of borrowing money lent at interest.

> Blockchain and cryptocurrencies enable massively distributed collaboration via **decentralized autonomous organizations** (DAOs) and other alternatives to traditional incorporation or partnership.

Is a future like that even possible? It depends who you ask. A legacy economist might scoff, but a renegade economist such as Kate Raworth would say we have no choice. Raworth is part of a new movement in economics that's questioning whether growth is truly the solution to all our problems, or if it might be time to transform our economic OS. In her 2018 TED Talk, she addressed traditional economic theory head on. "Twentieth-century economics assured us that if growth creates inequality, don't try to redistribute, because more growth will even things up again. If growth creates pollution, don't try to regulate, because more

growth will clean things up again. Except, it turns out, it doesn't, and it won't." Growth has delivered extraordinary benefits to humanity, but at a cost. The question is: Can we lift *everyone* up without destroying the planet? And even if we can, what then? GDP growth can't continue forever with finite resources.

> A new type of thinking is essential if mankind is to survive and move toward higher levels.
>
> —Albert Einstein

Of course, this is not a new idea. In 1968 R. Buckminster Fuller warned us about this in his masterwork *Operating Manual for Spaceship Earth*. Decades earlier, philosopher Bertrand Russell offered a damning critique of our approach to abundance in his essay "In Praise of Idleness."

> Suppose that, at a given moment, a certain number of people are engaged in the manufacture of pins. They make as many pins as the world needs, working (say) eight hours a day. Someone makes an invention by which the same number of men can make twice as many pins: pins are already so cheap that hardly any more will be bought at a lower price. In a sensible world, everybody concerned in the manufacturing of pins would take to working four hours instead of eight, and everything else would go on as before. But in the actual world this would be thought demoralizing. The men still work eight hours, there are too many pins, some employers go bankrupt, and half the men previously concerned in making pins are thrown out of work. There is, in the end, just as much leisure as on the other plan, but half the men are totally idle while half are still overworked. In this way, it is insured that the unavoidable leisure shall cause misery all round instead of being a universal source of happiness. Can anything more insane be imagined?

The message has been clear for decades. We have to move from a model of eternal consumption (a chart that goes up and to the right

forever) to a model that is regenerative and distributive. That doesn't mean regressive socialism or state-controlled enterprise; it means embodying new goals. The problem isn't free markets; it's what we value. It's the stories we tell ourselves. And unfortunately, we're stuck in a loop with our economic, educational, and social operating systems. They shape us. And we make (and remake) them in our own image. To break free we need new structures and platforms built to nurture and sustain Evolutionary Organizations. Luckily, they are starting to emerge.

Building Blocks

Building and sustaining an Evolutionary Organization is an uphill battle when the economic OS is structured to promote growth and gains over purpose and meaning. From incorporation to funding, every piece of the business ecosystem fits together. These structural impediments are entrenched and self-reinforcing. Which is why it's so difficult and important to create credible alternatives. Fortunately, the foundations of a new economic OS are starting to emerge, granting founders and shareholders the opportunity to change their underlying constraints.

> There is nothing more difficult to take in hand, more perilous to conduct, or more uncertain in its success, than to take the lead in the introduction of a new order of things.
>
> —Niccolò Machiavelli

New Forms of Incorporation

Officers in a traditional corporation have a legal obligation, known as a fiduciary duty, to act in the best interest of their shareholders. Of course, in practice, "best interest" translates to *maximizing shareholder return*. The law completely sidesteps any externalities—environmental or social impacts that might occur as a result of pursuing this goal. Barring any

regulations that stand in the way, corporate leadership is *obligated* to pollute the environment if it will provide immediate gains to share-holders, even if those shareholders will eventually drink that water or breathe that air. When it comes to their legal mandate, there's no room for purpose beyond profit.

But that's all about to change. Enter the *public benefit corporation*, a relatively new form of incorporation that allows an organization to pri-oritize a purpose or mission—beyond profit—that provides a benefit to society. By encoding this right and responsibility into the legal structure of the business itself, the public benefit corporation allows an organiza-tion to protect and nurture its mission, even as a publicly traded com-pany. This fills a gap that has long existed between for-profit and nonprofit organizations. For-profits achieve far greater scale, but they lack accountability for impact. Nonprofits are purpose driven, but they lack the means and motivation to engage in critical fields such as tech-nology, energy, healthcare, transportation, and infrastructure in a meaningful way. The public benefit corporation promises the best of both worlds: purpose *and* profit, scale *and* impact. As of this writing, thirty-four U.S. states have already passed legislation enabling this form of incorporation, and more are following suit. Some of the most progres-sive organizations around have already taken the leap and become pub-lic benefit corporations, including Method, Kickstarter, Plum Organics, Patagonia, and Danone North America. Good company.

While public benefit corporations represent a legal innovation, Certi-fied B Corporations represent an operational one. B Lab, the nonprofit that developed and promotes the model legislation for public benefit corpora-tions, is also the creator of a set of standards for social impact that make it more tangible and measurable. The idea here is to move from intentions to outcomes. Not just what an organization values, but what it does. To become a Certified B Corporation, an organization must take the B Impact Assessment, which measures impact areas such as governance, workers,

community, and environment, from the perspective of both operations and the underlying business model. Applicants that achieve an impact score above a certain threshold are certified, authorized to use the Certified B Corporation language and logo, and join a community of more than 5,997 organizations that represent 86 countries and more than 150 industries. Interestingly, many companies leverage both the public benefit corporation legal structure and the certification as a way to increase transparency and accountability in their pursuit of a new kind of performance.

Meanwhile, it's possible that one part of the future of incorporation is actually a blast from the past. **Cooperatives**, which are jointly owned and democratically controlled organizations, have existed for hundreds of years, with the earliest examples emerging in the late eighteenth and early nineteenth centuries. Co-ops generally embody a set of ideals known as the Rochdale Principles: (1) voluntary and open membership, (2) democratic member control, (3) economic participation by members, (4) autonomy and independence, (5) education, training, and information, (6) cooperation among cooperatives, and (7) concern for community. These manifest in an organization that is essentially *for the people, by the people*. While the legal structure of cooperatives is progressive, their approach to governance can be somewhat tedious. The "one member, one vote" principle can give way to the tyranny of consensus on every issue, large and small. That's why savvy cooperatives are beginning to experiment with models of self-management that use consent and elections to distribute authority and increase agility. And yet one study of cooperatives in Western Europe, the United States, and Latin America found that worker cooperatives are already *more* productive than conventional businesses.

And what if your mission is to create an organization that isn't incorporated or centralized at all? That ambition is at the heart of an emerging class of *decentralized autonomous organizations* (DAOs), which leverage blockchain and cryptocurrencies to create products and ser-

vices without central control. The best explanation of this controversial new form of organization comes from *Cointelegraph*, an independent online publication that covers the future of money. "Imagine a vending machine that not only takes money from you and gives you a snack in return but also uses that money to automatically re-order the goods. This machine also orders cleaning services and pays its rent all by itself. Moreover, as you put money into that machine, you and its other users have a say in what snacks it will order and how often it should be cleaned. It has no managers, all of those processes were pre-written into code." Developers, leveraging what they have learned in creating cryptocurrencies such as Bitcoin and Ethereum, are pioneering a new generation of decentralized applications that allow organizations to operate like that magical vending machine. Through a series of rules called smart contracts, founders can create, fund, and operate an entire organization independent of hierarchical management. Everything, from paying contributors for their work to making decisions about investment, is managed in a distributed way. Looking for the CEO? Headquarters? You won't find them. It's early days for this technology and the jury is still out on issues of security, how decisions are made, and the legal liability of contributors. But something tells me we haven't heard the last of DAOs.

New Forms of Investment

Concurrent with incorporation, every organization faces another early conundrum: how to raise the funds necessary to pursue its mission. While some businesses can be self-funded, a significant number of organizations require investment to get started and, later, to scale. Every company that takes on traditional investment is (often unwittingly) promising to grow forever in exchange for capital now. While it can be hard to imagine in the early days, the very nature of growth beyond scale

is exploitative. Take Google, the paragon of iconoclastic virtue when it rose to prominence with its "Don't be evil" slogan. You can't "organize the world's information" without inadvertently (or intentionally) *controlling* the flow of that information. Or take Facebook, the social network whose noble mission is to "bring the world closer together." As an ad-supported business, its growth depends on capturing more and more of our attention, which means it must—at a very fundamental cognitive level—manipulate and addict us. These are companies with seemingly virtuous founders who still have board-level control of their creations. Imagine what happens when the people at the reins are just hired guns.

Even firms that just *borrow* money are on the treadmill. The Federal Reserve System and privately held commercial banks create and lend money *at interest*, which creates a cascade of expected return for all lenders. Thus, businesses that borrow must recover that initial capital *and then some* in order to break even. Interestingly, at different points in the last century some economists have flirted with the idea of *negative interest*—a demurrage charge—as a way to encourage the flow of money in society. If you think back to the original forms of value storage—bread, grain, fruit, livestock—they all decayed over time. Sitting on a mountain of grain for years wouldn't make you rich; it would all go to waste and you'd be left with nothing. Instead you'd lend it out, at 0 percent interest or even less, in the hopes that you'd be paid back later when you needed it. In this model, surplus becomes valuable when it moves. When the same dollar is spent ten times in a community rather than just once, prosperity follows. We're a long way from deploying this at scale, but in times of crisis such as the Great Depression, communities have launched similar concepts with great success. Unsurprisingly, these experiments were quickly shuttered by the governments presiding over the recessions that triggered them.

In this century, for new forms of incorporation that prioritize purpose and public benefit to work, we need new investors and investment

vehicles that align with our values. Rising to meet that challenge, *impact investing* reflects a growing desire among investors to generate returns through investments that provide a public benefit. As concepts such as corporate social responsibility and triple bottom line (social, environmental, economic) have become more well known, the market has grown into a $250 billion industry on its way to half a trillion or more in the next decade. Private equity firm TPG recently launched the Rise Fund, the largest fund of its kind thus far, with more than $2 billion in committed capital and a founder's board that includes U2 frontman Bono, producer/philanthropist Jeff Skoll, Richard Branson, and Reid Hoffman. Even Bain Capital is in on the fun, with its $390 million Double Impact Fund.

Venture capital is also experiencing an awakening. Firms such as Better Ventures, Impact Engine, and Notley Ventures are putting their money where their mouth is without compromising their ability to generate significant returns. New York–based Social Impact Capital thinks beyond its investment to consider the robustness of its portfolio companies. In an interview with *Crunchbase*, managing partner Sarah Cone said, "One way I test [their concept] before investing is to ask myself, 'What if the most evil company in the world acquired this company? Could they end the social good component without killing the business entirely?' If the answer is yes, then we wouldn't invest." What a powerful filter.

And what about those organizations that need to go public to realize their vision? They face a situation so unappealing that today's hottest startups are avoiding an initial public offering (IPO) for as long as they possibly can. The average number of companies who choose an IPO each year is one third what it was prior to 2000. Uber and Airbnb were valued at more than $69 billion and $100 billion respectively when they went public. By comparison, when Amazon went public it had a market value of just $438 million. When companies wait to go public, only elite investors get to enjoy the climb. Yet all that's waiting for leadership on the other side of the IPO rainbow is the constant scrutiny and pressure of a

bunch of spectators. The CEO who seeks to invest in long-term success (or, gasp, purpose and impact) is met with outcries from activist investors, while algorithmic high-frequency traders create and profit from short-term volatility.

Eric Ries, the best-selling author of *The Lean Startup* and *The Startup Way*, has a better idea. A much better idea. Over the last five years he's been quietly developing an alternative called the Long-Term Stock Exchange (LTSE). There are several key differences between the LTSE and a traditional stock exchange. First, shareholder voting power is based on tenure. The longer you hold a stock, the more influence you'll have. Second, the exchange requires that listed companies have specific ties between executive pay and long-term performance. And finally, additional disclosure requirements increase transparency in both directions between shareholders and management. The net effect? Organizations that go public on the LTSE will be set up for what Airbnb cofounder Brian Chesky calls an "infinite time horizon." No more trading the future for the quarter, my friends. We're in this for the long haul. Eric's renegade startup is learning and pivoting in precisely the way you'd expect, given his pedigree. But we can rest a little easier knowing that a shot has been fired across the bow of the stock market status quo.

Just the Beginning

These innovations are not *solutions*. They are the start of a renaissance in the form and function of organizations. Our job isn't simply to adopt them blindly but to push them, advocate for them, and evolve them into serious alternatives that can stand on their own merit. Today less than 1 percent of all organizations are experimenting with these new possibilities. The tipping point is far off in the distance. It is people like you who will draw this future closer and make the impossible possible for the next generation of founders and teams.

Into the Great Wide Open

Many unanswered questions remain when it comes to the future of work. So many possibilities lie before us, we may find ourselves breathlessly optimistic one moment and bitterly cynical the next. To conclude our journey together, I want to address a few of the questions that provoke me the most. No easy answers exist. Just strong opinions loosely held. My hope is that by raising these topics now, I can set you on a path to answering them more deeply for yourself in the months and years to come.

Can everyone work this way? What about people who lack the requisite intelligence or maturity? It's easy to get caught up in the idea that self-management requires a certain level of intellect and sophistication. But I think that's a mistake. It requires emotional maturity and competence, both of which can be nurtured in *anyone* in the right environment. Not everyone can be a rocket scientist, but not everyone can cook pit barbecue, either. Our natural talents, education, and socialization create diversity, which in the right context can be a source of power. I've seen self-management positively flourish in places where workers historically have been treated as disposable—in factories, in fast food, in retail—and these cases give me hope that the future of work can include everyone. An old Henry Ford quote best sums up my feelings on this question: "Whether you believe you can do a thing or not, you are right."

Will there be enough work for everyone? If organizations get smaller and technology grows ever more capable, what will we all do all day? Artificial intelligence, robotics, and automation are reshaping the workforce as we speak. When Marc Andreessen said that "software is eating the world," he wasn't kidding. Everywhere you look, the complicated work is becoming the domain of technology. The bots in my company already do the work of a couple people—handling everything from collecting feedback to scheduling meetings. Self-driving trucks such as the one Tesla is developing threaten the livelihoods of many

more. Some 3.5 million people make their living driving trucks in the United States alone. A staggering 39 percent of the jobs in the legal sector could be gone in the next ten years. What happens next depends on us. If we continue to take Milton Friedman's advice, that the purpose of business is solely to generate returns for shareholders, we're going to preside over a spectacular and unintentional recession, unlike anything we've ever seen. Because, to put it bluntly, guess what happens when a significant chunk of the population is unemployed? They can't buy *your* stuff. And on and on.

But here's the good news. Even as complicated work goes away, there's a whole world of complex work yet to be done. Researchers have concluded that nonroutine work—anything that does not conform to checklists and rules—is far less likely to be automated in the years to come. Drones can water the vines, but they can't run the winery. An app can help you meditate, but it can't do couples counseling. An AI can design a good workout, but it's a lot easier to blow off your iPhone than a real trainer waiting for you at the gym. Technology cannot invent the future, because that is the domain of people. It will augment us. It will change the shape of our teams. It will change the paths of our careers. But we will still have work to do. And that work will be more creative, more complex, and more challenging than anything we've ever faced.

What about education? Do we have to change how we develop young minds in order to prepare for this new world of work? Absolutely. After all, Taylorism didn't stop with the workplace. It also completely infected education in order to prepare generations of workers for the industrial jobs they would one day inherit. The values and structures present in education today are mirror images of the Legacy OS we're trying to over-throw. Teachers (bosses) have authority; students don't. High school seniors whom we trust to drive on our streets need a bathroom pass to use the toilet. The curriculum (the work) is fixed, not fluid or self-

directed, and often woefully outdated. A million books are published every year and kids are still reading *The Scarlet Letter*. Learning happens through lecture (instruction) rather than curiosity and experience. Subjects (functions) are isolated rather than integrated. Math is pure math. Science is pure science. Use math to create something? Forget about it. The measure of mastery is a test (training and certification), not practice or the ability to deliver outcomes. Failure is punished. Compliance is rewarded. It's staggeringly similar to the modern workplace, isn't it? Similar and equally irrelevant. The educational revolutionary Sugata Mitra said as much in his 2013 TED Prize acceptance speech: "[We] engineered a system so robust that it's still with us today, continually producing identical people for a machine that no longer exists." But there is hope on the horizon. Several schools, including Sugata Mitra's School in the Cloud, the Evangelical School Berlin Centre (ESBC) in Germany, and Ricardo Semler's Lumiar in Brazil are radically reinventing education for the twenty-first century. Gone are traditional grades, lectures, and subjects, replaced with an integrated approach that honors active participation and curiosity. Everyone—students, parents, and faculty—is expected to participate in the creation of the curriculum. These institutions are cracking the code on education. The question is, can enough schools unlearn the habits of the past to produce a generation ready to take us into the future?

What about cultures that still value hierarchy, compliance, or other legacy traits? Do entire countries and peoples have to change to be successful? Isn't that colonialism? I routinely encounter this line of questioning when I deliver a speech outside the United States. And it's totally fair. Ideas such as autonomy, inclusion, transparency, and radical candor are perceived very differently by different cultures. To generalize for a moment based on the research of INSEAD professor Erin Meyer, cultures vary wildly on a whole host of styles, from how they lead to how they communicate. Japanese business culture, for example, is very hier-

archical. And yet it employs a more consensual approach to decision making, in which proposals are circulated for support at the lower levels before being presented to leadership. When it comes to negative feedback, Israelis might be perceived as more direct, while Indonesians prefer an indirect approach. It would be inaccurate to say that some cultures are ahead and others are behind. Rather, every culture has ingrained strengths and weaknesses when it comes to work. What is incumbent on every business leader, regardless of cultural context, is to ask, What about our way of working is serving us and what isn't? We can celebrate our cultures and diversity, and we can rise above them. What is critical is that we not cling to traditions or norms without a conscious choice to accept the trade-offs they contain.

If every organization is People Positive and Complexity Conscious, how will they differentiate themselves? If purpose, sustainability, and human centricity are pervasive, what does that mean for competition? This is, to me, the most interesting question of all. These things are differentiators today, but tomorrow they could be table stakes. If every grocery store chain were as progressive as Whole Foods, how would I decide where to shop? If every clothing company were as conscious as Patagonia or Everlane, how would I decide what to wear? If every workplace were a "best place to work," how would I choose where to apply? It's an interesting thought experiment, and I think it's the shadow hope of every participant in this movement—that the world will catch up. But here's the thing: human beings are limited only by their imagination. Before Tony Hawk successfully landed the first 900—a two-and-a-half-revolution aerial spin on a skateboard—the move was thought impossible. Thirteen years later, twelve-year-old Tom Schaar completed a 1080. We raise the bar by seeing what is possible. That's why I think the answer to this question is that organizations will differentiate themselves by doing *more*. You think you're customer obsessed? Look at Amazon. You think you're sustainable? Just wait until you see what Patagonia does

next. The point is this: what a *good problem to have*. As the team at B Lab says, if everyone were competing not to be the best *in* the world, but to be the best *for* the world, how incredible that would be. The promise of capitalist competition, but with a new scoreboard. Instead of just one winner, everybody wins.

As we learned at the beginning, all models are wrong, but some are useful. So let me say this before we depart: this book is not perfect. Mistakes were made. Perhaps I portrayed an organization differently from how you (or the organization) would. Perhaps I failed to describe a concept or distinction as well as another expert might. It's even possible that something *big* is missing—something that will reveal itself to us in years to come and irrevocably transform how we organize. I can live with that. Because what I've presented here is *enough*. It's enough to stoke that disenchantment that you feel. That this can't be our best. That we're capable of more. It's enough to get started. Enough to go on. And that's all that matters. Progress over perfection. Courage over caution. This isn't business as usual. This is brave new work.

Acknowledgments

First and foremost, a heartfelt thank you to my amazing wife and best friend, Britt. Somehow she managed to move us across the country (in the midst of this project) so flawlessly that we never missed a beat. She is the very definition of grace. In related news, five-year-olds are not often patient, but I'm grateful to my son, Huxley, for being so understanding and supportive throughout this process. The funny thing is, we all ended up spending more time together than we usually do, and I loved every minute of it.

I come from a family of people who are passionate about their work. My dad, Pete, is my original thought partner, so I'm lucky that he cares as much about this subject as I do. He and his wife, Kelly, helped me break through mental barriers more than once. My mother, Barb, is the source of my confidence and dedication. Watching her buck the bureaucracy of the school system for years was early inspiration. Hats off to my brother, Ben, for having the courage to stop doing work he didn't love. And a sincere thanks to the rest of my clan, including my wife's wonderfully eccentric family, my indefatigable grandmother, and the most interesting collection of aunts, uncles, and cousins you could possibly imagine.

I've never met a group that so epitomizes "clear eyes, full hearts, can't lose" as the team at Portfolio. An animated unicorn GIF goes to my

editor, Leah Trouwborst, without whom this book would not exist. She saw the potential before anyone else did. Adrian Zackheim and Will Weisser deciphered the book's message like literary code breakers. Their commitment to getting it right meant everything to me. And to the rest of our intrepid crew, Tara Gilbride, Alie Coolidge, Helen Healey, Ryan Boyle, Daniel Lagin, Chris Sergio, Eric White, Matthew Boezi, and Hilary Roberts goes a windmill high five. Without you this is just the world's longest blog post.

Unending gratitude goes out to my team—the most stunning colleagues anyone could ever ask for. You have helped me grow in so many directions. To quote *Hamilton*, "Tomorrow there'll be more of us."

Cheers to all my former colleagues over the years, particularly those who were around when I was first tinkering with this stuff. You jumped in with me and I'll never forget that. To those of you out there "changing how the world works" in your own way, I salute you.

Such an expansive topic required a fairly unusual research process. Azy Groth and Tim Casasola dunked on it every day for months. Jared Lindzon and Aimee Groth also played their part in the early days. Champagne for my real friend Meg Thompson, longtime literary agent and publishing spirit guide. You stuck with me for years while this percolated. Now we dance! And a tip of the hat to James Robinson, my trusty speaking agent, who finds excuses to send me all around the world to tell this story. Couldn't do it without you, old sport.

The Ready wouldn't exist without brave clients, past and present. Your commitment to an ever better future inspires me. And sometimes a leader comes along and changes us as much as we change them. Appreciation to Beth Comstock, Linda Boff, Sarah Wills, Raghu Krishnamoorthy, Russell Stokes, Susan Sobbott, Scott Roen, Frank Cooper, Simon Lowden, Massimo Portincaso, Scott Harrison, Lauren Letta, Marc Lien, John Polstein, and Lorin Thomas Tavel for those moments. A secret

handshake goes out to all the catalysts, right-hand women and men, influencers, iconoclasts, and rebels who found a way to invite us into their organizations. Without you the world stands still.

When you're as passionate about work as I am, friends are hard to come by and harder to keep. That's why I'm blown away by Ben Kaufman, Rachel Shechtman, Ricky Van Veen, Steve Roberts, Steve Holt, Eliot Drake, Brian Swibel, and Scott Belsky. You each inspire me with your commitment to saying and doing what matters.

I can say with confidence that very few of the ideas in this book originated with me. I owe a debt to the many brave and brilliant people who have pioneered these ways of working and shaped my thinking about organizations. A complete list of influences would be far too long (you should see my *Brave New Work* library), so I'll simply thank those who have taught me the most: Dennis Bakke, Steve Blank, Jos de Blok, Bjarte Bosgnes, Jacob Bøtter, Brian Carney, James Carse, W. Edwards Deming, David Dewane, Peter Drucker, Amy Edmondson, Charles Eisenstein, Gerard Endenburg, Robin Fraser, Jason Fried, Isaac Getz, James Gleick, Seth Godin, Deborah Gordon, Paul Graham, Adam Grant, Dave Gray, Gary Hamel, David Heinemeier Hansson, Tim Harford, Frederick Herzberg, Jeremy Hope, Steven Johnson, Daniel Kahneman, Kevin Kelly, David Kidder, Doug Kirkpatrick, Henrik Kniberg, Lars Kolind, John Kotter, Frederic Laloux, Jason Little, David Marquet, John E. Mayfield, Douglas McGregor, Greg McKeown, Melanie Mitchell, Taiichi Ohno, Tom Peters, Niels Pflaeging, Daniel Pink, Adam Pisoni, Eric Ries, Brian Robertson, Ricardo Semler, Peter Senge, Simon Sinek, Dave Snowden, Nassim Taleb, Ben Thompson, Geoffrey West, Meg Wheatley, Keith Yamashita, Jean-Francois Zobrist, and the few I forgot. Especially them, because their work has stuck with me but their names have not. I also owe a special thank-you to Tom Thomison, who introduced me to many of these ideas at just the right time, starting a fire that surprised us both.

It's funny how forty-eight hours with someone can be the start of so much. And of course, big thanks to Douglas Rushkoff for being my accountability buddy as we penned our most recent manifestos.

Finally, I'd like to express my admiration for the countless teams out there where the sidewalk ends, inventing the future of work. May you inspire more people to join you there.

Appendix

EVOLUTIONARY ORGANIZATIONS

In order to discover a diverse and proven collection of organizations working in new ways, I used the following criteria to guide my search:

1. **The organization must be larger than ten people.** While this may seem arbitrary, I have found that the commitment to new ways of working is tested when an organization begins to scale and confronts its own complexity.

2. **The organization must be at least five years old.** Longevity affords an organization time to develop an OS that is richer and more distinctive.

3. **The organization must have documented principles and practices that reflect a People Positive and Complexity Conscious way of working.** These practices do not need to be evenly distributed across the entire culture, but they do need to be firmly established and flourishing where present.

These criteria led me and my research team to consider hundreds of organizations from around the globe, both past and present (historical cases were considered and included). The organizations listed are the

ones that inspired us most. While these firms operate with significantly more individual autonomy, they also utilize deeply thoughtful processes and social norms to guide behavior. Rather than top-down control, they leverage positive peer pressure and transparency to encourage responsible and coordinated behavior. They separate the complex from the complicated and leverage the forgotten art of dialogue to create empathy and learning in place of mindless execution. The absence of traditional bosses doesn't make these organizations leaderless, it makes them leader*ful*.

Let me save you the suspense, though: these organizations are not perfect. Every one of these cultures has its struggles and tensions. Some have found success in one function or location but not another. Some have issues around diversity and inclusion. Some are falling victim to their own scale and success, reverting to bureaucracy when deep down they know better. But here's the thing: they're working on it. The firms on this list ensure that the mechanisms to turn tension into change are present for every member every day.

I'll also point out that I spent months deliberating about whether to include juggernauts such as Google, Amazon, and Facebook on this list. They excel at learning and adapting in an almost unprecedented way, but they sometimes struggle to remain human, and their aggregate impact on the world concerns me. While they are often far ahead of the mean in terms of autonomy and transparency, they have a long way to go compared to many of the organizations on this list. With that in mind, I decided that while I would highlight some of their most admirable practices in the text, I would not characterize them as Evolutionary Organizations at this time. Therefore, below you'll find two lists: Evolutionary Organizations—cultures that met the criteria above—and Sources of Inspiration—cultures that I deemed too nascent or uneven in their commitment but with much to teach us nonetheless.

If there's one thing I've learned in my travels it's that the world is

a big place. In the course of writing this book I've been introduced to dozens of cases and cultures that deserve to be recognized, and I'm convinced that there are hundreds (or even thousands) more out there. So if you know of, or participate in an Evolutionary Organization, I invite you to visit BraveNewWork.com and share your story with the rest of us.

Evolutionary Organizations

AES	Haier
Askinosie Chocolate	Handelsbanken
Basecamp	Haufe-umantis
Black Lives Matter	Heiligenfeld
Blinkist	Hengeler Mueller
Bridgewater	Herman Miller
Buffer	HolacracyOne
Burning Man	Ian Martin Group / Fitzii
Buurtzorg	Incentro
BvdV	John Lewis
charity: water	Joint Special Operations Command
Crisp	Kickstarter
David Allen Company	Lumiar Schools
dm-drogerie markt	Menlo Innovations
elbdudler	Mondragon
Endenburg Elektrotechniek	Morning Star
Enspiral	Nearsoft
Equinor	Netflix
Evangelical School Berlin Centre	Nucor
FAVI	Orpheus Chamber Orchestra
Gini	Patagonia
GitLab	Phelps Agency
Gumroad	Pixar

Premium-Cola

Promon Group

Red Hat

School in the Cloud

Schuberg Philis

Semco Group

Spotify

stok

Sun Hydraulics

Treehouse

USS *Santa Fe*

Valve

Whole Foods

W. L. Gore

WP Haton

Zalando Technology

Zappos

Zingerman's

Sources of Inspiration

Airbnb

Amazon

Automattic

Chipotle

Chobani

Danone North America

Etsy

Everlane

Facebook

GitHub

Google

Johnsonville

Lyft

Medium

Quicken Loans

Slack

Southwest Airlines

Stack Overflow

Toyota

Warby Parker

Wikimedia

Zapier

USING THE OS CANVAS

The canvas can provoke incredible conversations and powerful stories. It can help you and your team identify what to amplify and what to change. It can even help you find unexpected sources of inspiration. But for your first foray into what can be an emotional and challenging conversation, we recommend a lightly structured workshop format that has

proven to be both safe and effective. Here are some brief instructions that will help you on your way.

1. **Setup.** Reserve a quiet and spacious room with at least one big blank wall or window. Invite your team or a group of no more than fifteen interested participants from across the company. Print out each of the twelve OS Canvas dimensions (available at BraveNew Work.com) and place them on the wall prior to the session. Provide two pads of sticky notes (yellow and green) and a Sharpie for each participant.

2. **Check-in.** Ask each participant to "check in" to the meeting by answering the following question: What has your attention? Push the group for vulnerability and candor. Model it by checking in first.

3. **Introduction.** Introduce participants to the concept of the organizational operating system through selected passages from this book, and facilitate a brief discussion.

4. **Tensions.** Ask everyone to generate a list of tensions they're feeling day to day based on the following questions: What is preventing you (or your team) from doing your best work? What is slowing you (or your team) down? What is our biggest problem as an organization? Ask each attendee to write down each of their tensions on a yellow sticky note (one per note), with a goal of at least five per person.

5. **Bright spots.** Ask everyone to generate a list of the things that are going well based on the following questions: What is working? What is enabling you to do your best work? What is speeding you (or your team) up? What is helping you make better decisions? What are you proud of in terms of the way we work? What is our biggest strength as an organization? Ask each attendee to write

down each bright spot on a green sticky note (one per note), with a goal of at least five per person.

6. **Placement.** Now have each person place their tensions and bright spots on the canvas where they feel the issue is rooted most strongly. As they place each one, have them offer a short explanation to the group. Allow the notion that some tensions belong in multiple dimensions. Allow questions and answers.

7. **Discussion.** One dimension at a time, discuss the tensions that were placed on the canvas and identify themes and patterns. As a group, ask yourselves: Why are these issues present now? What underlying structures, rules, or processes are contributing to them? What personal behavior, attitudes, and assumptions are contributing to them? How are the tensions in different dimensions connected? Are there bigger patterns present in the canvas (e.g., we have limited trust, so we reserve all decisions for management, so we have lots of meetings with people asking for permission)?

8. **Sensing.** Based on everything you've all heard, ask the group what they would like to change. What one thing would they flip on its head? What addition or subtraction might unlock other possibilities? Capture any intentions or next actions and use them as fodder for your first few loops. For a refresher on looping see page 201.

Notes

PART ONE: THE FUTURE OF WORK

7 **destabilize their own communities:** William J. Donovan, *Simple Sabotage Field Manual* (Washington, D.C.: Oxford City Press, 2011).

8 **Operating Organization of the Union Pacific:** Ray Morris, *Operating Organization of the Union Pacific and Southern Pacific Systems, Railroad Administration* (New York and London: Appleton, 1920), Baker Old Class Collection, Baker Library Historical Collections, Harvard Business School.

9 **With an estimated 311,000 signals:** National Transportation Operations Coalition, "2012 National Traffic Signal Report Card: Technical Repot," www.ite.org /pub/?id=e265477a-2354-d714-5147-870dfac0e294.

11 **one for every 1,118 intersections:** John Metcalfe, "Why Does America Hate Roundabouts?" *CityLab*, March 10, 2016, www.citylab.com/transportation/2016 /03/america-traffic-roundabouts-street-map/408598.

11 **work normally during power outages:** "Roundabout Benefits," Washington State Department of Transportation, accessed September 1, 2018, www.wsdot .wa.gov/Safety/roundabouts/benefits.htm.

12 **"There's no word for accountability":** Anu Partanen, "What Americans Keep Ignoring About Finland's School Success," *The Atlantic*, December 29, 2011, www.theatlantic.com/national/archive/2011/12/what-americans-keep-ignoring -about-finlands-school-success-250564.

20 **"Stop trying to borrow wisdom":** Jason Yip, "Japan Lean Study Mission Day 4: Toyota Home, Norman Bodek, and Takeshi Kawabe," *You'd think with all my video game experience that I'd be more prepared for this: Agile, Lean, Kanban* (blog), December 8, 2008, http://jchyip.blogspot.com/2008/12/japan-lean-study -mission-day-4-toyota.html.

21 **"Hardly a competent workman":** Frederick Winslow Taylor, "Shop Management," paper presented at the American Society of Mechanical Engineers meeting in Saratoga, New York, June 1903.

22 **"All you have to do is take orders":** Robert Kanigel, *The One Best Way: Frederick Winslow Taylor and the Enigma of Efficiency* (Cambridge, MA: MIT Press, 2005), 214.

22 **"your way of doing the job for mine":** Robert Kanigel, "Taylor-Made. (19th-Century Efficiency Expert Frederick Taylor)," *The Sciences* 37, no. 3 (May 1997): 1–5.

23 **bestselling business books of the decade:** Tim Hindle, *Guide to Management Ideas and Gurus* (London: John Wiley & Sons, 2008).

23 **They develop a science:** Frederick Winslow Taylor, *Principles of Scientific Management* (New York and London: Harper & Brothers, 1911), 36–37.

25 **"going to do in future periods":** Duff McDonald, *The Firm* (New York: Oneworld, 2014).

26 **you'll get your toilet paper:** NateIBEW558l, comment on "[Serious] Reddit, What's the Worst Example of Workplace Bureaucracy You've Ever Encountered?" Reddit, r/AskReddit, October 18, 2014, www.reddit.com/r/AskReddit /comments/2jmvro/serious_reddit_whats_the_worst_example_of.

26 **the word itself is the unlikely pairing:** *Merriam-Webster Dictionary Online*, s.v. "Bureaucracy," accessed May 26, 2013, www.merriam-webster.com/dictionary /bureaucracy; "-cracy," accessed September 12, 2018, www.merriam-webster .com/dictionary/-cracy.

26 **23.8 million management roles:** Gary Hamel and Michele Zanini, "Excess Management Is Costing the U.S. $3 Trillion Per Year," *Harvard Business Review*, September 5, 2016, https://hbr.org/2016/09/excess-management-is-costing-the -us-3-trillion-per-year.

27 **16 percent of our working lives:** Hamel makes this inference about U.S. workers based on Australia data, "Get Out of Your Own Way: Unleashing Productivity" (Building the Lucky Country: Business Imperatives for a Prosperous Australia, no. 4), Deloitte, 2014, https://www2.deloitte.com/au/en/pages/building -lucky-country/articles/get-out-of-your-own-way.html.

27 **"in the early stages of a startup":** Steve Blank, "Organizational Debt Is Like Technical Debt—but Worse," *Steve Blank* (blog), May 19, 2015, https://steveblank .com/2015/05/19/organizational-debt-is-like-technical-debt-but-worse.

29 **the average global citizen:** Max Roser, "Economic Growth," Our World in Data, 2018, https://ourworldindata.org/economic-growth.

29 **sixfold increase in prosperity:** Max Roser, "The Short History of Global Living Conditions and Why It Matters That We Know It," Our World in Data, 2018, https://ourworldindata.org/a-history-of-global-living-conditions-in-5-charts.

29 **by 2027 the average tenure:** Innosight, "Creative Destruction Whips Through Corporate America" (executive briefing), Winter 2012, www.innosight.com/wp -content/uploads/2016/08/creative-destruction-whips-through-corporate -america_final2015.pdf.

29 **will shrink to just twelve years:** Scott D. Anthony, S. Patrick Viguerie, and Andrew Waldeck, "Corporate Longevity: Turbulence Ahead for Large Organizations," Innosight, Spring 2016, www.innosight.com/wp-content/uploads /2016/08/Corporate-Longevity-2016-Final.pdf; Scott D. Anthony et al., "2018 Corporate Longevity Forecast: Creative Destruction Is Accelerating" (executive briefing), Innosight, February 2018, www.innosight.com/wp-content/uploads /2017/11/Innosight-Corporate-Longevity-2018.pdf.

29 *all firms* **was roughly 10.5 years:** Madeleine I. G. Daepp et al., "The Mortality of

Companies," *Journal of the Royal Society Interface* 12, no. 106 (2015), doi: 10.1098 /rsif.2015.0120.

30 **returns depend on *volatility*:** Evelyn Cheng, "Just 10% of Trading Is Regular Stock Picking, JPMorgan Estimates," *CNBC,* June 13, 2017, www.cnbc.com /2017/06/13/death-of-the-human-investor-just-10-percent-of-trading-is -regular-stock-picking-jpmorgan-estimates.html.

30 **"the market is a voting machine":** Benjamin Graham as told to Warren Buffett. Recounted in Berkshire Hathaway's 1987 letter to shareholders, available here: www.berkshirehathaway.com/letters/1987.html.

31 **ROA decline from 12.9 percent:** Wassili Bertoen and Maarten Oonk, "The Big Shift," Deloitte, January 6, 2017, https://www2.deloitte.com/nl/nl/pages/center -for-the-edge/artikelen/the-big-shift.html.

31 **in 1965 to 8.3 percent in 2015:** John Hagel et al., "Success or Struggle: ROA as a True Measure of Business Performance," Deloitte, October 30, 2013, https: //www2.deloitte.com/insights/us/en/topics/operations/success-or-struggle-roa -as-a-true-measure-of-business-performance.html.

32 **mergers fail to provide the revenue:** Scott A. Christofferson, Robert S. McNish, and Diane L. Sias, "Where Mergers Go Wrong," *McKinsey Quarterly,* May 2004, www.mckinsey.com/business-functions/strategy-and-corporate-finance/our -insights/where-mergers-go-wrong.

32 **83 percent of the deals had actually failed:** John Kelly, Colin Cook, and Don Spitzer, "Unlocking Shareholder Value: The Keys to Success," KPMG, 1999, http://people.stern.nyu.edu/adamodar/pdfiles/eqnotes/KPMGM&A.pdf.

33 **34 has been declining since 1990:** "Frequently Asked Questions About Small Business," U.S. Small Business Administration, Office of Advocacy, August 2017, www.sba.gov/sites/default/files/advocacy/SB-FAQ-2017-WEB.pdf.

33 **researchers started tracking this data:** J. D. Harris, "The Decline of American Entrepreneurship—in Five Charts," *The Washington Post,* February 12, 2015, www.washingtonpost.com/news/on-small-business/wp/2015/02/12/the-decline -of-american-entrepreneurship-in-five-charts/?utm_term=.c65afa6f0d87.

33 **an entrepreneurial boom:** Ben Casselman, "The Slow Death of American Entrepreneurship," *FiveThirtyEight,* May 15, 2014, https://fivethirtyeight.com /features/the-slow-death-of-american-entrepreneurship.

33 **in the space of two decades:** Ian Hathaway and Robert Litan, "The Other Aging of America: The Increasing Dominance of Older Firms," Brookings Institution, July 2014, www.brookings.edu/wp-content/uploads/2016/06/other _aging_america_dominance_older_firms_hathaway_litan.pdf.

33 **sixteen times more patents:** "Frequently Asked Questions About Small Business," U.S. Small Business Administration, September 2012, www.sba.gov/sites /default/files/FAQ_Sept_2012.pdf.

33 **In 2016 it was actually negative:** U.S. Bureau of Labor Statistics, "Multifactor productivity trends, 2017," March 21, 2018, www.bls.gov/news.release/prod3 .nr0.htm; Shawn Sprague, "Below Trend: The U.S. Productivity Slowdown Since the Great Recession," *Beyond the Numbers: Productivity* 6, no. 2 (January 2017), www.bls.gov/opub/btn/volume-6/below-trend-the-us-productivity-slowdown -since-the-great-recession.htm.

34 **the ratio of pay between the CEO:** Lawrence Mishel and Jessica Schneider, "CEO Pay Remains High Relative to the Pay of Typical Workers and High Wage Earners," Economic Policy Institute, July 20, 2017, www.epi.org/publication/ceo-pay-remains-high-relative-to-the-pay-of-typical-workers-and-high-wage-earners.

34 **from 22.3:1 to 271:1:** Mishel and Alyssa Davis, "Top CEOs Make 300 Times More Than Typical Workers," Economic Policy Institute, June 21, 2015, www.epi.org/publication/top-ceos-make-300-times-more-than-workers-pay-growth-surpasses-market-gains-and-the-rest-of-the-0-1-percent.

35 **Buurtzorg has been able to avoid:** Claire Lickman, "Humanity Over Bureaucracy: New Models of Care, by Alieke Van Dijken," *The Happy Manifesto* (blog), June 27, 2016, https://happymanifesto.com/2016/06/27/full-talk-humanity-bureaucracy-alieke-van-dijken; "Buurtzorg's Healthcare Revolution: 14,000 Employees, 0 Managers, Sky-High Engagement," *Corporate Rebels* (blog), June 21, 2017, https://corporate-rebels.com/buurtzorg.

35 **leaders and functions for more than ten years:** Frederic Laloux, *Reinventing Organizations* (Brussels, Belgium: Nelson Parker, 2014).

38 **European holdout in its space:** Brian M. Carney and Isaac Getz, *Freedom, Inc.: How Corporate Liberation Unleashes Employee Potential and Business Performance* (New York: Somme Valley House, 2016).

38 **FAVI exports *to* China:** Matthew E. May, "Mastering the Art of Bosslessness," *Fast Company*, September 26, 2012, www.fastcompany.com/3001574/mastering-art-bosslessness.

38 **Rooted in the work of psychologists:** Abraham H. Maslow, *Maslow on Management* (Homewood, IL: R. D. Irwin, 1965); Carl R. Rogers, *On Becoming a Person: A Therapist's View of Psychotherapy* (Boston: Houghton Mifflin, 1961).

39 **"There may be said to be two classes":** Brené Brown, *Daring Greatly: How the Courage to Be Vulnerable Transforms the Way We Live, Love, Parent, and Lead* (New York: Gotham, 2012), 142.

39 **The average human being has an inherent dislike of work:** Douglas McGregor, *The Human Side of Enterprise: Annotated Edition*, commentary by Joel Cutcher-Gershenfeld (New York: McGraw Hill, 2006), 4.

40 **The expenditure of physical and mental effort:** Ibid.

41 **He asked students:** Daniel H. Pink, *Drive: The Surprising Truth About What Motivates Us* (New York: Riverhead Books, 2009), 5–9.

42 **control-oriented firms:** Ibid., 89; Myrlande Davermann, "Why HR = Higher Revenues," *CNN Money*, September 22, 2006, https://money.cnn.com/magazines/fsb/fsb_archive/2006/07/01/8380516/index.htm?postversion=2006092217.

43 **keep temporary workers longer:** "The Story of FAVI: The Company That Believes That Man Is Good," European Workplace Innovation Network, http://uk.ukwon.eu/File%20Storage/5160692_7_The-story-of-favi.pdf.

43 **almost 1,000 percent since 2004:** Google Trends, s.v. "VUCA," accessed September 1, 2018, https://trends.google.com/trends/explore?date=all&q=VUCA.

47 **1978 has used anything but the flop:** 12th IAAF World Championships in Athletics, "IAAF Statistics Handbook, Berlin 2009" (Monte Carlo: IAAF Media & Public Relations Department, 2009), 554–55.

48 **Studies show you probably won't:** William Samuelson and Richard Zeck-

hauser, "Status Quo Bias in Decision Making," *Journal of Risk and Uncetainty* 1 (1988), 7–59, https://sites.hks.harvard.edu/fs/rzeckhau/status%20quo%20bias.pdf.

48 **The same is true:** James J. Choi, David Laibson, and Brigitte C. Madrian, "Plan Design and 401(k) Savings Outcomes," working paper 10486 (Cambridge, MA.: National Bureau of Economic Research, 2004): www.nber.org/papers /w10486.pdf.

48 **We prefer to stick with what we've got:** Daniel Kahneman, Jack L. Knetsch, and Richard H. Thaler, "Anomalies: The Endowment Effect, Loss Aversion, and Status Quo Bias," *Journal of Economic Perspectives* 5, no. 1 (1991), doi:10.1257/jep.5.1.193.

49 **"put a man on the moon":** Astro Teller, "Google X Head on Moonshots: 10x Is Easier Than 10 Percent," *Wired*, February 11, 2013, www.wired.com/2013/02 /moonshots-matter-heres-how-to-make-them-happen.

49 **"All models are wrong":** George E. P. Box and Norman R. Draper, *Empirical Model-Building and Response Surfaces* (New York: John Wiley & Sons, 1987), 440.

PART TWO: THE OPERATING SYSTEM

56 **human-centric design principles:** Gary Hamel, "First, Let's Fire All the Managers," *Harvard Business Review*, December 2011, https://hbr.org/2011/12/first -lets-fire-all-the-managers.

59 **"business is to increase its profits":** Milton Friedman, "The Social Responsibility of Business Is to Increase Its Profits," *The New York Times Magazine*, September 13, 1970, www.colorado.edu/studentgroups/libertarians/issues/friedman-soc -resp-business.html.

59 *result* **of success is uninspiring:** Scott Winship, "What Really Happened to Income Inequality in the 20th Century?" *The Atlantic*, May 14, 2012, www .theatlantic.com/business/archive/2012/05/what-really-happened-to-income -inequality-in-the-20th-century/257156.

59 **"Saying that the purpose of a company":** Kevin Laws, "Successful Startups Don't Make Money Their Primary Mission," *Harvard Business Review*, July 10, 2015, https://hbr.org/2015/07/successful-startups-dont-make-money-their -primary-mission.

60 **after the publication of the book:** "Performance of Firms of Endearment," Firms of Endearment, accessed September 1, 2018, www.firmsofendearment.com.

60 **ceases to be a good measure:** Marilyn Strathern, "'Improving Ratings': Audit in the British University System," *European Review* 5, no. 3 (July 1997): 305–321.

62 **"both meaningful and measurable":** Greg McKeown, *Essentialism: The Disciplined Pursuit of Less* (New York: Crown Business, 2014), 126.

63 **tighten up your road map:** Ben Barry, "Facebook's Little Red Book," Office of Ben Barry, accessed September 1, 2018, http://v1.benbarry.com/project/facebooks -little-red-book.

66 **internet-famous employee handbook:** Valve, *Handbook for New Employees* (Bellevue, WA: Valve Press, 2012), www.valvesoftware.com/company/Valve _Handbook_LowRes.pdf.

67 **retention, operations, and promotions:** "Our Story," David Marquet, accessed September 1, 2018, www.davidmarquet.com/our-story.

68 **"Policies are organizational scar tissue"**: Jason Fried and David Heinemeier Hansson, *Rework* (New York: Crown Business, 2010), 260.

69 **authority structures is W. L. Gore:** "Our Beliefs & Principles," Gore, accessed September 1, 2018, www.gore.com/about/our-beliefs-and-principles.

76 **"leaders and members of their teams"**: Bill Fischer, Umberto Lago, and Fang Liu, "The Haier Road to Growth," *strategy+business*, April 27, 2015, www.strategy-business.com/article/00323?gko=c8c2a.

76 **leveraging its unconventional structure:** Zhang Ruimin, "Why Haier Is Reorganizing Itself Around the Internet of Things," *strategy+business*, February 26, 2018, www.strategy-business.com/article/Why-Haier-Is-Reorganizing-Itself-around-the-Internet-of-Things?gko=895fe.

78 **speed, learning, and collaboration:** Niels Pflaeging, "Org Physics: How a Triad of Structures Allows Companies to Absorb Complexity," LinkedIn, February 12, 2017, www.linkedin.com/pulse/org-physics-how-triad-structures-allows-companies-absorb-pflaeging.

84 **4.5 million hours reading content:** Ken Yueng, "Medium Grows 140% to 60 Million Monthly Visitors," *Venture Beat*, December 14, 2016, https://venturebeat.com/2016/12/14/medium-grows-140-to-60-million-monthly-visitors.

84 **"Upon further reflection"**: Ev Williams, "Renewing Medium's Focus," *3 Min Read* (blog), *Medium*, January 4, 2017, https://blog.medium.com/renewing-mediums-focus-98f374a960be.

85 **If it succeeds:** Rob Reid, "Medium Targets 10 Million Paying Members in 5 Years," *NewCo Shift* (blog), *Medium*, October 13, 2017, https://shift.newco.co/medium-targets-10-million-paying-members-in-5-years-40029cf2bb8f.

85 **"Don't buy this jacket"**: Patagonia, "Don't Buy This Jacket, Black Friday and the New York Times," *The Cleanest Line* (blog), *The Cleanest Line*, November 25, 2011, www.patagonia.com/blog/2011/11/dont-buy-this-jacket-black-friday-and-the-new-york-times.

87 **"People with targets"**: Daryl Kulak and Hong Li, *The Journey to Enterprise Agility: Systems Thinking and Organizational Legacy* (Cham, Switzerland: Springer International Publishing, 2017), 172.

88 **"situations that have already changed"**: Adrian Cho, *The Jazz Process: Collaboration, Innovation, and Agility, Portable Documents* (Boston: Pearson, 2010), 94.

88 **deployment every second:** Wermer Vogels, "The Story of Apollo—Amazon's Deployment Engine," *All Thing Distributed* (blog), November 12, 2014, www.allthingsdistributed.com/2014/11/apollo-amazon-deployment-engine.html.

88 **"get a good test of an experience"**: Andrea Huspeni, "Why Mark Zuckerberg Runs 10,000 Facebook Versions a Day," *Entrepreneur*, accessed September 1, 2018, www.entrepreneur.com/article/294242.

89 **sure the trade-offs are real:** Jordan Husney, "Strategic Prioritization Using 'Even Over' Statements," *Parabol* (blog), *Medium*, October 22, 2017, https://focus.parabol.co/strategic-prioritization-using-even-over-statements-fb63e78e7b4d.

90 **And you're ready to act:** Dave Snowden, "Think Anew, Act Anew: Scenario Planning," *Dave Snowden* (blog), *Cognitive Edge*, April 21, 2009, http://cognitive-edge.com/blog/think-anew-act-anew-scenario-planning.

94 **"case it will be dangerous"**: Jeremy Hope and Robin Fraser, "Who Needs Bud-

gets?" *Harvard Business Review*, February 2003, https://hbr.org/2003/02/who
-needs-budgets.

94 **bank over the last ten years:** Niels Kroner, *A Blueprint for Better Banking:
Svenska Handelsbanken and a Proven Model for More Stable and Profitable Bank-
ing* (Hampshire, UK: Harriman, 2011).

96 **$1.2 billion a year:** Hope and Fraser, "Who Needs Budgets?"

96 **"Budgets also sub-optimize performance":** Steve Morlidge, *The Little Book of
Beyond Budgeting* (Leicestershire, UK: Matador, 2017), 19.

98 **"The strength of polycentric governance":** John Tierney, "The Non-Tragedy of
the Commons," *TierneyLab* (blog), *The New York Times*, October 15, 2009, https://
tierneylab.blogs.nytimes.com/2009/10/15/the-non-tragedy-of-the-commons.

99 **budgets that persist year after year:** David Kesmodel, "Meet the Father
of Zero-Based Budgeting," *The Wall Street Journal*, March 26, 2015, www.wsj
.com/articles/meet-the-father-of-zero-based-budgeting-1427415074.

100 **whole new level of engagement:** Cobudget, accessed September 1, 2018, https://
cobudget.co/#; Participatory Budget Project, accessed September 1, 2018, www
.participatorybudgeting.org; "Enspiral Handbook," Enspiral, accessed September
1, 2018, https://handbook.enspiral.com.

103 **beyond its initial advantage:** Wynne Parry, "Exaptation: How Evolution Uses
What's Available," *Live Science*, September 16, 2013, www.livescience.com/39688
-exaptation.html.

103 **cleaning product for wallpaper:** Sam Biddle, "Whoops! The 10 Greatest
(Accidental) Inventions of All Time," *Gizmodo*, August 26, 2010, https://gizmodo
.com/5620910/whoops-the-10-greatest-accidental-inventions-of-all-time.

105 **"easily distinguishable operationally":** Justin Fox, "Amazon, the Biggest R&D
Spender, Does Not Believe in R&D," *Bloomberg*, April 12, 2018, www.bloomberg
.com/view/articles/2018-04-12/amazon-doesn-t-believe-in-research-and
-development-spending.

106 **look for the next big thing:** Andrew J. Smart, "Why Organizations Should
Embrace Randomness Like Ant Colonies," *Harvard Business Review*, September
13, 2013, https://hbr.org/2013/09/why-organizations-should-embra.

108 **how might we validate that:** "The Lean Startup Methodology," The Lean
Startup, accessed September 1, 2018, http://theleanstartup.com/principles.

111 **two hours and thirty minutes:** "Dec 01: 1913: Ford's Assembly Line Starts
Rolling," This Day in History, History.com, December 1, 2009, www.history.com
/this-day-in-history/fords-assembly-line-starts-rolling.

112 **1,500 different companies around the world:** Sumesh Krishnan and Dr. Mukul
Shukla, *Concepts in Engineering Design* (Chennai, India: Notion Press, 2016), 6.

113 **May we all be so intentional:** Henrik Kniberg, "Spotify Engineering Culture
(Part 1)," *Spotify Labs* (blog), March 27, 2014, https://labs.spotify.com/2014/03/27
/spotify-engineering-culture-part-1.

115 **"think we can ship by Friday":** John Zeratsky, "Sprints Are the Secret to Getting
More Done," *Harvard Business Review*, March 15, 2016, https://hbr.org/2016/03
/sprints-are-the-secret-to-getting-more-done.

119 **$37 billion for U.S. businesses:** Michael Mankins, Chris Brahm, and Greg Caimi,
"Your Scarcest Resource," *Harvard Business Review*, May 2014, https://hbr.org/2014

/05/your-scarcest-resource; "You Waste a Lot of Time at Work," Atlassian, accessed September 1, 2018, www.atlassian.com/time-wasting-at-work-infographic.

119 **"20 people and no one notices":** Vala Afshar (@ValaAfshar), "You likely have to get management approval for a $500 expense ... but you can call a 1 hour meeting with 20 people and no one notices," Twitter, July 8, 2015, 9:59 A.M., https://twitter .com/valaafshar/status/618826783799025664?lang=en.

119 **unnecessary in the future of work:** Eugene Kim, "Slack, the $2.8 Billion Startup That Wants to Kill Email, Really Is Reducing Work Email," *Business Insider*, October 29, 2015, www.businessinsider.com/slack-survey-shows-it-reduces-work -email-2015-10.

120 **"The film—not the filmmaker":** Ed Catmull, "Inside The Pixar Braintrust," *Fast Company*, March 12, 2014, www.fastcompany.com/3027135/inside-the-pixar -braintrust.

120 **unearth the greatness within:** Ed Catmull, "How Pixar Fosters Collective Creativity," *Harvard Business Review*, September 2008, https://hbr.org/2008/09 /how-pixar-fosters-collective-creativity; "Pixar," Box Office Mojo, www.box officemojo.com/franchises/chart/?id=pixar.htm.

120 **gatherings of their global membership:** Oliver Staley, "The Creator of WordPress Shares His Secret to Running the Ultimate Remote Workplace," *Quartz at Work*, May 29, 2018, https://work.qz.com/1289444/automattics-secret -to-successful-remote-work-is-having-everyone-meet-in-person.

125 **instructions for how to facilitate them:** Henri Lipmanowicz and Keith McCand-less, *The Surprising Power of Liberating Structures: Simple Rules to Unleash a Cul-ture of Innovation* (Seattle, WA: Liberating Structures Press, 2014); "Liberating Structures," accessed September 1, 2018, www.liberatingstructures.com.

129 **the leader of al-Qaeda in Iraq:** Stanley McChrystal, "The Military Case for Sharing Knowledge," TED video, 6:44, March 2014, www.ted.com/talks/stanley _mcchrystal_the_military_case_for_sharing_knowledge/transcript#t-218264.

129 **"a system in which large networks":** Melanie Mitchell, *Complexity: A Guided Tour* (New York: Oxford University Press, 2009), 13.

130 **individual and organizational learning:** Harold Jarche, "Organize for Com-plexity," *Harold Jarche* (blog), May 1, 2014, http://jarche.com/2014/05/organize -for-complexity.

131 **gigabytes of information *every day*:** Harris Andrea, "The Human Brain Is Loaded Daily with 34 GB of Information," *Tech 21 Century* (blog), accessed Sep-tember 1, 2018, www.tech21century.com/the-human-brain-is-loaded-daily-with -34-gb-of-information.

134–35 **304 emails per week:** The Radicati Group, Inc., "Email Statistics Report, 2017–2021," February 2017, www.radicati.com/wp/wp-content/uploads/2017/01 /Email-Statistics-Report-2017-2021-Executive-Summary.pdf; "You Waste a Lot of Time at Work," Atlassian, accessed September 1, 2018, www.atlassian.com /time-wasting-at-work-infographic.

135 **213 percent ROI over three years:** Kate Borger, "The Total Economic Impact™ of G Suite," Google, February 2016, https://gsuite.google.com/learn-more/office -cloud-comparison.html.

135 **interrogated by the masses:** Rob Walker, "How Reddit's Ask Me Anything

Became Part of the Mainstream Media Circuit," *Yahoo! News*, March 13, 2013, www.yahoo.com/news/how-reddit-s-ask-me-anything-became-part-of-the-mainstream-media-circuit--130755591.html.

142 **"each person in the Lattice":** W. L. Gore & Associates, Inc., "The Lattice Organization," https://folk.uio.no/terjegro/materials/Gore_lattice.pdf.

142 **Organizational psychologist Adam Grant:** Stanford eCorner, "Adam Grant: Hire for Culture Fit or Add?" YouTube video, 4:19, February 16, 2017, www.youtube.com/watch?v=mLHp25mUd40.

143 **In fact, a recent McKinsey study:** Vivian Hunt, Dennis Layton, and Sara Prince, "Why diversity matters," McKinsey & Company, January 2015, www.mckinsey.com/business-functions/organization/our-insights/why-diversity-matters.

143 **think: soldier or firefighter:** Francesca Gino and Michael I. Norton, "Why Rituals Work," *Scientific American*, May 14, 2013, www.scientificamerican.com/article/why-rituals-work.

144 **Amazon, which acquired Zappos:** Ruth Umoh, "Why Amazon pays employees $5,000 to quit," CNBC, May 21, 2018, www.cnbc.com/2018/05/21/why-amazon-pays-employees-5000-to-quit.html.

147 **what makes you tick:** Adam Bryant, "Want to Know Me? Just Read My User Manual," Corner Office, *The New York Times*, March 30, 2013, www.nytimes.com/2013/03/31/business/questbacks-lead-strategist-on-his-user-manual.html; David Politis, "This Is How You Revolutionize the Way Your Team Works Together . . . and All It Takes Is 15 Minutes," LinkedIn, March 29, 2016, www.linkedin.com/pulse/how-you-revolutionize-way-your-team-works-together-all-david-politis.

148 **avoid it because it's uncomfortable:** Ayla Lewis, "Gratitude in the Workplace: Research-Based Tools to Increase Happiness and Engagement," Happy Brain Science, accessed September 1, 2018, www.happybrainscience.com/blog/5-workplace-gratitude-tools.

153 **"hiding their limitations":** Robert Kegan and Lisa Laskow Lahey, *An Everyone Culture: Becoming a Deliberately Developmental Organization* (Boston: Harvard Business Review Press, 2016), 1.

153 **"continually clarifying and deepening our personal vision":** Peter M. Senge, *The Fifth Discipline: The Art & Practice of The Learning Organization* (New York: Currency, 1990), 7.

153 **"embrace reality and deal with it":** Robert Kegan et al., "Making Business Personal," *Harvard Business Review*, April 2014, https://hbr.org/2014/04/making-business-personal.

154 **satisfaction and job performance:** Timothy A. Judge, Edwin A. Locke, Cathy C. Durham, Avraham N. Kluger, "Dispositional effects on job and life satisfaction: The role of core evaluations," *Journal of Applied Psychology* 83, no.1 (February 1998): 17–34, doi:10.1037/0021-9010.83.1.17, PMID 9494439.

154 **other people shape their lives:** Julian B. Rotter, "Generalized expectancies for internal versus external control of reinforcement," *Psychological Monographs: General and Applied* 80, no. 1 (1966): 1–28, doi:10.1037/h0092976.

154 **Carol Dweck later expanded this concept:** Carol S. Dweck, *Mindset: The New Psychology of Success* (New York: Random House, 2006).

157 **say more than we can write down:** Dave Snowden, "Rendering Knowledge, *Dave Snowden* (blog), Cognitive Edge, October 11, 2008, http://cognitive-edge .com/blog/rendering-knowledge.

157 **grow in- and outside of work:** Michelle Traub, "About Us: Etsy School," *Etsy News* (blog), August 23, 2013, https://blog.etsy.com/news/2013/about-us-etsy-school.

158 **doing appraisals annually:** Peter Cappelli and Anna Tavis, "The Performance Management Revolution," *Harvard Business Review*, October 2016, https://hbr .org/2016/10/the-performance-management-revolution.

164 **Latin word for salt:** Peter Gainsford, "Salt and Salary: Were Roman Soldiers Paid in Salt?" *Kiwi Hellenist* (blog), January 11, 2017, http://kiwihellenist.blogspot .com/2017/01/salt-and-salary.html.

164 *correlation* **between pay and results:** Alfie Kohn, "Why Incentive Plans Cannot Work," *Harvard Business Review*, September–October 1993, https://hbr.org/1993 /09/why-incentive-plans-cannot-work.

165 **any meaningful or lasting way:** Frederick Herzberg, Bernard Mausner, Barbara B. Snyderman, *The Motivation to Work*, 3rd ed. (New York: John Wiley & Sons, 1959), 42.

165 **income were correlated with happiness:** Daniel Kahneman and Angus Deaton, "High Income Improves Evaluation of Life but Not Emotional Well-being," *Proceedings of the National Academy of Sciences of the United States* 107, no. 38 (September 2010), doi: 10.1073/pnas.1011492107.

165 **"money buys you misery":** Phyllis Korkki, "Job Satisfaction vs. a Big Pay-check," *The New York Times*, September 11, 2010, www.nytimes.com/2010/09/12 /jobs/12search.html.

166 **privilege than with pay scale:** Claire Cain Miller, "Children Hurt Women's Earnings, but Not Men's (Even in Scandinavia)," *The New York Times*, February 5, 2018, www.nytimes.com/2018/02/05/upshot/even-in-family-friendly-scandinavia -mothers-are-paid-less.html; Sarah Kliff, "A Stunning Chart Shows the True Cause of the Gender Wage Gap," *Vox*, February 19, 2018, www.vox.com/2018 /2/19/17018380/gender-wage-gap-childcare-penalty; Judith Warner and Dan-ielle Corley, "The Women's Leadership Gap," Center for American Progress, May 21, 2017, www.americanprogress.org/issues/women/reports/2017/05/21/432758 /womens-leadership-gap.

167 **salary you'd earn if you worked there:** "Buffer's Transparent Salary Calcula-tor," Buffer, accessed September 1, 2018, https://buffer.com/salary/software -engineer-mobile/average; "Calculate Your Salary," Stack Overflow, accessed September 1, 2018, https://stackoverflow.com/jobs/salary.

168 **"title 'Director of Engineering'":** Reed Hastings, "Netflix Culture: Freedom and Responsibility," SlideShare, August 1, 2009, www.slideshare.net/reed2001 /culture-1798664.

169 **deck to invent the future:** Stacey Leasca, "These Are the Highest Paying Jobs in the Gig Economy," *Forbes*, July 17, 2017, www.forbes.com/sites/sleasca/2017 /07/17/highest-paying-jobs-gig-economy-lyft-taskrabbit-airbnb/#e2d9eeb7b644.

170 **wage policy and productivity:** MIT Press, summary of *Wage Dispersion: Why Are Similar Workers Paid Differently?* by Dale T. Mortensen, accessed September 1, 2018, https://mitpress.mit.edu/books/wage-dispersion.

174 **But it simply isn't so:** Joshua Isaac Smith, "The Myth of Rational Decisions—How Emotions Flavour Our Choices," Adapt Faster, May 17, 2017, www.adapt faster.com/the-myth-of-rational-decisions-how-emotions-flavour-our-choices.

PART THREE: THE CHANGE

180 **"People like us do things":** Seth Godin, "People like us do things like this," Seth's Blog, July 26, 2013, https://seths.blog/2013/07/people-like-us-do-stuff-like-this.

180 **"Culture is like a shadow":** Niels Pflaeging (@NielsPflaeging), "Org #culture is like a shadow: You cannot change it, but it changes all the time. Culture is read-only." Twitter, March 25, 2015, 9:53 P.M., https://twitter.com/nielspflaeging/status/580955646415245312.

186 **really be done concurrently and continuously:** John P. Kotter, *Accelerate: Building Strategic Agility for a Faster-Moving World* (Boston: Harvard Business Review Press, 2014).

188 **"Managing the present to actually create":** AcademiWales, "Dave Snowden - How leaders change culture through small actions," Youtube video, 1:11:06, July 26, 2016, https://youtu.be/MsLmjoAp_Dg?t=2m56s.

189 **"present can reinvent itself":** Steven Johnson, "The Genius of the Tinkerer," *The Wall Street Journal*, September 25, 2010, www.wsj.com/articles/SB100014240 5274870398930457550373010186083?mod=WSJ_hp_mostpop_read.

189 **"we manage the emergence":** CognitiveEdge, "How to Organise a Children's Party," YouTube video, 2:58, October 27, 2009, https://youtu.be/Miwb92eZaJg?t=2m32s.

189 **"milk in coffee":** Niels Pflaeging, "Change is more like adding milk to coffee," *Medium*, March 6, 2018, https://medium.com/@NielsPflaeging/change-is-more-like-adding-milk-to-coffee-b6498a3aa708

192 **"Do this again next year":** Ed Catmull, *Creativity, Inc.: Overcoming the Unseen Forces That Stand in the Way of True Inspiration* (New York: Random House, 2014), 174.

196 **"the past and the future":** Dave Gray, *Liminal Thinking: Create the Change You Want by Changing the Way You Think* (Brooklyn: Two Waves Books, 2016), xxi.

202 **"What does tension seek?":** Peter M. Senge, *The Fifth Discipline: The Art & Practice of the Learning Organization* (New York: Currency, 1990), 140.

202 **"gap between how things *are*":** Brian Robertson, *Holacracy* (New York: Macmillan, 2015), 5.

217 **fifty-three words for snow:** David Robson, "There Really Are 50 Eskimo Words for 'Snow,'" *The Washington Post*, January 14, 2013, www.washingtonpost.com/national/health-science/there-really-are-50-eskimo-words-for-snow/2013/01/14/e0e3f4e0-59a0-11e2-beee-6e38f5215402_story.html?utm_term=.506583668442.

218 **network of eighteen coaches:** "Buurtzorg's Healthcare Revolution: 14,000 Employees, 0 Managers, Sky-High Engagement," *Corporate Rebels* (blog), June 21, 2017, https://corporate-rebels.com/buurtzorg.

221 **"shared belief held by members":** Amy Edmondson, "Psychological Safety and

Learning Behavior in Work Teams," *Administrative Science Quarterly* 44, no. 2 (June 1999): 350–383.

222 **confide in one another:** Alexandra Jamieson and Bob Gower, *Getting to Hell Yes* (self-published, 2018), gettingtohellyes.com.

227–28 **"All managers at Handelsbanken":** Handelsbanken, "Anders Bouvin, new President and Group Chief Executive of Handelsbanken," press release, August 16, 2016, http://news.cision.com/handelsbanken/r/anders-bouvin--new-president-and-group-chief-executive-of-handelsbanken,c2059556.

230 **"do things that don't scale":** Paul Graham, "Do Things That Don't Scale," *Paul Graham*, http://paulgraham.com/ds.html.

236 **recent research by David DeSteno:** David DeSteno, *The Truth About Trust: How It Determines Success in Life, Love, Learning, and More* (New York: Plume, 2014).

237 **"Questions are places in your mind":** Jason Fried, "What are questions?," *Medium*, August 1, 2016, https://m.signalvnoise.com/what-are-questions-51c20fde777d.

EPILOGUE: WHAT DREAMS MAY COME

245 **"if you're giving back, you took too much":** Ricardo Semler, "How to run a company with (almost no rules)," TED video, 21:43, October, 2014, www.ted.com/talks/ricardo_semler_how_to_run_a_company_with_almost_no_rules.

246 **"Twentieth-century economics assured us":** Kate Raworth, "A healthy economy should be designed to thrive, not grow," TED video, 15:53, April 2018, www.ted.com/talks/kate_raworth_a_healthy_economy_should_be_designed_to_thrive_not_grow.

247 **"Can anything more insane be imagined?":** Bertrand Russell, *In Praise of Idleness, and Other Essays* (London: Routledge, 2004), 6, www.zpub.com/notes/idle.html.

250 **known as the Rochdale Principles:** "The Rochdale Principles," Co-operative Heritage Trust, accessed September 1, 2018, www.rochdalepioneersmuseum.coop/about-us/the-rochdale-principles.

250 **one study of cooperatives:** Virginie Pérotin, "What Do We Really Know About Worker Co-operatives?" Co-operatives UK, no date, www.uk.coop/resources/what-do-we-really-know-about-worker-co-operatives.

251 **"pre-written into code":** "What Is DAO," *Cointelegraph*, accessed July 31, 2016, https://cointelegraph.com/ethereum-for-beginners/what-is-dao#how-daos-work.

252 **becomes valuable when it moves:** Charles Eisenstein, *Sacred Economics: Money, Gift, and Society in the Age of Transition* (Berkeley, CA: North Atlantic Books, 2011).

253 **producer/philanthropist Jeff Skoll:** "A New Global Impact Fund," The Rise Fund, accessed September 1, 2018, http://therisefund.com.

253 **$390 million Double Impact Fund:** "Scaling Mission-Driven Companies," Bain Capital Double Impact, accessed September 1, 2018, www.baincapitaldoubleimpact.com.

253 **"If the answer is yes, then we wouldn't invest":** Mary Ann Azevedo, "Growth with an Impact: The Rise of VCs Looking to Fund a (Profitable) Cause," *Crunchbase*, February 2, 2018, https://news.crunchbase.com/news/growth-impact-rise-vcs-looking-fund-profitable-cause.

254 **"real" businesses bleed black:** Indie VC, "We Invest in Real Businesses," accessed September 1, 2018, www.indie.vc.

254 **The average number of companies:** Adrian Wooldridge, "America can't control the global flow of ideas," *The Economist*, September 13, 2018, www.economist .com/business/2017/04/22/why-the-decline-in-the-number-of-listed-american -firms-matters.

254 **value of just $438 million:** Alex Wilhelm, "A Look Back in IPO: Amazon's 1997 Move," *TechCrunch*, June 28, 2017, https://techcrunch.com/2017/06/28/a-look-back -at-amazons-1997-ipo.

256 **"software is eating the world":** Marc Andreessen, "Why Software Is Eating the World," *The Wall Street Journal*, August 20, 2011, www.wsj.com/articles /SB10001424053111903480904576512250915629460.

256 **gone in the next ten years:** Michael Grothaus, "Bet You Didn't See This Coming: 10 Jobs That Will Be Replaced by Robots," *Fast Company*, January 19, 2017, www .fastcompany.com/3067279/you-didnt-see-this-coming-10-jobs-that-will-be -replaced-by-robots.

257 **"[We] engineered a system so robust":** Sugata Mitra, "Build a school in the cloud," TED video, 22:25, February 2013, www.ted.com/talks/sugata_mitra_build _a_school_in_the_cloud.

Index

PENGUIN PARTNERSHIPS

Penguin Partnerships is the Creative Sales and Promotions team at Penguin Random House. We have a long history of working with clients on a wide variety of briefs, specializing in brand promotions, bespoke publishing and retail exclusives, plus corporate, entertainment and media partnerships.

We can respond quickly to briefs and specialize in repurposing books and content for sales promotions, for use as incentives and retail exclusives as well as creating content for new books in collaboration with our partners as part of branded book relationships.

Equally if you'd simply like to buy a bulk quantity of one of our existing books at a special discount, we can help with that too. Our books can make excellent corporate or employee gifts.

Special editions, including personalized covers, excerpts of existing books or books with corporate logos can be created in large quantities for special needs.

We can work within your budget to deliver whatever you want, however you want it.

For more information, please contact
salesenquiries@penguinrandomhouse.co.uk